AF882547

It’s New To Me!

100 iPhone Tips for Everyday Users

Melvin Coates

Renegade Publishings 2026

Copyright

Published by Renegade Publishings

ISBN: 978-1-956088-15-1

https://link.content360.io

Contents

Getting Started

Tip 1: Use 3D Touch

3D Touch is a pressure-sensitive feature available on certain iPhone models that allows you to access additional options and shortcuts by pressing firmly on the screen. It enhances navigation and interaction by providing quick actions without opening apps fully. This feature can speed up common tasks and improve overall efficiency.

Watch Tutorial:
Tap here
https://link.content360.io/iphone-3d-touch-quick-actions

Figure 1: Scan to Watch

To use 3D Touch:

1. Firmly press on an app icon on the Home screen to reveal quick action shortcuts.
2. Press on a message, email, or link to preview its content without fully opening it (known as "Peek").
3. Press a bit harder while previewing to open the content fully ("Pop").
4. Adjust the 3D Touch sensitivity in Settings > Accessibility > Touch > 3D & Haptic Touch to suit your preference.

A common use case for 3D Touch is managing emails. Instead of opening the Mail app and navigating to a message, press firmly on an email preview to peek at the content and decide if it requires immediate attention. If it does, press a little deeper to open it, or swipe up while peeking to access options like Mark as Read or Delete, saving time and effort.

Tip 2: Use The Reachability Feature

The Reachability feature on the iPhone is designed to make it easier to use the device with one hand by bringing the top of the screen closer to your thumb. This is especially helpful on larger iPhone models, where reaching the top of the screen can be challenging without adjusting your grip.

Watch Tutorial:
Tap here
https://link.content360.io/iphone-reachability-one-hand

Figure 2: Scan to Watch

To activate Reachability, lightly swipe down on the bottom edge of the screen (just above the Home indicator) on iPhones without a Home button. On models with a Home button, quickly double-tap (not press) the Home button. The screen will shift down, allowing you to access icons and controls near the top more comfortably. To return to the normal view, simply tap the empty area at the top or swipe up from the bottom again.

For example, if you are trying to tap a notification or a control in the top menu bar while holding your iPhone in one hand, Reachability can bring those elements within easy reach, reducing the chance of dropping your phone or having to adjust your grip awkwardly.

Tip 3: Use The Magnifier

The Magnifier is a built-in accessibility feature on the iPhone that transforms the device's camera into a digital magnifying glass. It allows you to zoom in on small text or objects, adjust brightness and contrast, and apply filters to improve readability or visibility. This tool is especially helpful for reading fine print, examining details, or viewing objects up close without needing a physical magnifying glass.

Watch Tutorial:
Tap here
https://link.content360.io/iphone-magnifier-accessibility-tool

Figure 3: Scan to Watch

To use the Magnifier on your iPhone:

1. Open the Settings app and tap on Accessibility.
2. Select Magnifier and toggle the switch to turn it on.
3. To quickly access the Magnifier, triple-click the side button (or the Home button on older models).
4. The Magnifier interface will open, showing a live camera view.
5. Use the slider to zoom in or out as needed.
6. Tap the filter icon to apply different color filters or adjust brightness and contrast.
7. Tap the flashlight icon to turn on the LED light for better illumination.
8. Move your iPhone closer to or farther from the object to get the desired focus and size.

Imagine you're at a restaurant and the menu has very small print. Instead of struggling to read it, simply activate the Magnifier, zoom in on the text, and adjust the brightness or apply a filter to enhance clarity. This makes it easier to see the menu items without needing additional tools or assistance.

Tip 4: Use The Calculator

The Calculator app on your iPhone is a handy tool for performing quick mathematical calculations without needing a separate device or software. It offers basic functions like addition, subtraction, multiplication, and division, as well as more advanced operations such as percentages and scientific calculations when the phone is in landscape mode.

Watch Tutorial:
Tap here
https://link.content360.io/iphone-calculator-hidden-gestures

Figure 4: Scan to Watch

To use the Calculator app:

1. Open the Calculator app by tapping its icon on the Home screen or in the Control Center.
2. Enter the first number using the number pad.
3. Tap the desired operation symbol (+, −, ×, ÷).
4. Enter the second number.
5. Tap the equals (=) button to see the result.
6. For more advanced functions, rotate your iPhone to landscape orientation to access additional scientific calculator features.
7. To clear the current calculation, tap the "C" button.

Use the Calculator app when you need to quickly add up expenses, convert measurements, calculate tips at a restaurant, or solve basic math problems on the go without opening a separate device or app.

Tip 5: Use Spotlight Search

Spotlight Search is a powerful tool on the iPhone that allows you to quickly find apps, contacts, emails, messages, and even web results without having to open multiple apps. It helps you save time by providing a central place to search for almost anything on your device or the internet.

Watch Tutorial:
Tap here
https://link.content360.io/spotlight_search

Figure 5: Scan to Watch

To use Spotlight Search: 1. From the Home Screen, swipe down in the middle of the screen. 2. Tap the search bar that appears at the top. 3. Type the keyword or phrase related to what you are looking for. 4. Review the search results, which may include apps, contacts, messages, emails, calendar events, and web suggestions. 5. Tap the desired result to open it directly.

For example, if you want to quickly call a friend but don't want to open the Phone app and scroll through contacts, you can swipe down on the Home Screen, type their name in Spotlight Search, and tap their contact to place the call immediately. This saves time and streamlines accessing information on your iPhone.

Tip 6: Use The Back Tap Feature

The Back Tap feature on the iPhone allows you to perform specific actions by simply tapping the back of your device. This accessibility tool can be customized to trigger functions like taking a screenshot, opening the Control Center, or launching an app, making it easier to interact with your phone without navigating through menus.

Watch Tutorial:
Tap here
https://link.content360.io/back_tap

Figure 6: Scan to Watch

To enable and customize Back Tap: 1. Open the Settings app on your iPhone. 2. Scroll down and tap on Accessibility. 3. Select Touch, then scroll to find and tap Back Tap. 4. Choose either Double Tap or Triple Tap. 5. From the list of available actions, select the one you want to assign to the chosen tap gesture. 6. Exit Settings; the feature is now active.

Use Back Tap to quickly capture a screenshot while reading an article or to instantly open the camera when you want to take a photo without fumbling through the home screen. This feature streamlines common tasks, saving time and effort.

Tip 7: Customize Your Home Screen

Customizing your iPhone home screen allows you to organize apps, widgets, and wallpapers in a way that suits your preferences and enhances your device's usability. By rearranging app icons, adding widgets, and selecting a personalized wallpaper, you can create a more efficient and visually appealing interface.

Watch Tutorial:
Tap here
https://link.content360.io/customize_home_screen

Figure 7: Scan to Watch

Steps: 1. Touch and hold any app icon on your home screen until the icons start to jiggle. 2. Drag apps to rearrange them or move them to different pages. 3. To create a folder, drag one app icon over another and release. 4. Tap the "+" button in the top-left corner to open the widget gallery. 5. Select a widget, choose its size, and tap "Add Widget" to place it on your home screen. 6. Tap and hold the widget to move it or adjust its position. 7. Press the home button or tap "Done" in the top-right corner (on devices without a home button, tap "Done") when finished. 8. To change the wallpaper, go to Settings > Wallpaper > Choose a New Wallpaper, and select an image from Apple's stock options or your photos.

Use Case: If you frequently use apps like Calendar, Weather, and Notes, adding their widgets to your home screen lets you view important information at a glance without opening each app. Organizing these apps into folders by category, such as Productivity or Entertainment, reduces clutter and makes navigation quicker, improving your overall iPhone experience.

Tip 8: Use Widgets On The Lock Screen

Widgets on the iPhone lock screen provide quick access to important information and app functions without unlocking the device. They offer a convenient way to glance at weather updates, calendar events, reminders, or fitness stats, enhancing productivity and personalization.

Watch Tutorial:
Tap here
https://link.content360.io/use_widgets_lock

Figure 8: Scan to Watch

To add widgets to the lock screen: 1. Wake your iPhone and stay on the lock screen. 2. Press and hold the lock screen until the customization menu appears. 3. Tap "Customize" below the lock screen preview. 4. Select the widget area beneath the clock. 5. Browse through available widgets and tap the ones you want to add. 6. Adjust or remove widgets by tapping the minus icon or dragging to reorder. 7. Tap "Done" to save your changes.

For example, if you have an important meeting, adding the Calendar widget to your lock screen lets you quickly check upcoming events without unlocking your phone. Similarly, fitness enthusiasts can monitor their activity rings or step count instantly, making it easier to stay motivated throughout the day.

Tip 9: Set Up Face ID

Face ID is a facial recognition system that allows you to unlock your iPhone, authorize purchases, and sign into apps quickly and securely using just your face. It uses advanced technology to create a detailed 3D map of your face, ensuring accurate recognition even in different lighting conditions or if you change your hairstyle.

Watch Tutorial:
Tap here
https://link.content360.io/setup_face_id

Figure 9: Scan to Watch

To set up Face ID on your iPhone: 1. Open the Settings app and tap on "Face ID & Passcode." 2. Enter your passcode when prompted. 3. Tap "Set Up Face ID." 4. Position your face inside the frame on the screen. 5. Slowly move your head in a circle to complete the first scan. 6. Tap "Continue" and repeat the head movement for the second scan. 7. Once complete, tap "Done" to finish the setup.

Using Face ID simplifies daily interactions with your iPhone. For example, when you receive a notification, you can glance at your phone to unlock it instantly without typing a passcode. It also speeds up purchases on the App Store or Apple Pay by authenticating with a quick facial scan, making transactions both convenient and secure.

Tip 10: Use The App Library

The App Library on your iPhone organizes all your apps into categorized folders, making it easier to find and manage them without cluttering your home screen. It automatically groups apps based on their functionality, such as Social, Productivity, or Entertainment, and provides a search bar to quickly locate any app.

Watch Tutorial:
Tap here
https://link.content360.io/app_library

Figure 10: Scan to Watch

To access the App Library, swipe left past your last home screen page. You can browse through the categorized folders or use the search bar at the top to type the name of the app you need. To open an app, simply tap on its icon. If you want to add an app from the App Library back to your home screen, press and hold the app icon, then select "Add to Home Screen."

Use the App Library when you want to keep your home screen organized and less crowded. It's especially helpful if you have many apps but prefer a minimalist home screen setup, allowing you to access any app quickly without scrolling through multiple pages.

Tip 11: Use Siri Shortcuts

Siri Shortcuts allow you to automate tasks and perform multiple actions with a single voice command or tap. By creating custom shortcuts, you can streamline daily routines, save time, and enhance your iPhone's functionality without needing to open multiple apps.

Watch Tutorial:
Tap here
https://link.content360.io/siri_shortcut

Figure 11: Scan to Watch

Steps: 1. Open the Shortcuts app on your iPhone. 2. Tap the "+" icon to create a new shortcut. 3. Select "Add Action" and choose from a wide range of available actions, such as sending a message, playing music, or setting an alarm. 4. Combine multiple actions to create a sequence that completes a task automatically. 5. Name your shortcut and customize its icon if desired. 6. Save the shortcut and add it to your Home Screen, widget, or assign a Siri phrase for voice activation.

Use Case: Imagine you want to create a shortcut that prepares you for a workout. You can build a shortcut that turns on Do Not Disturb, starts a workout playlist in Apple Music, and opens your preferred fitness app—all triggered by saying, "Hey Siri, start my workout." This saves you from manually adjusting settings and launching apps one by one.

Tip 12: Use Quick Settings From Lock Screen

Quick Settings on the iPhone allow you to access frequently used controls like Wi-Fi, Bluetooth, brightness, and volume without having to unlock your device or navigate through multiple menus. By accessing Quick Settings directly from the lock screen, you can make adjustments swiftly and efficiently, saving time and streamlining your interaction with the phone.

Watch Tutorial:
Tap here
https://link.content360.io/use_quick_settings

Figure 12: Scan to Watch

To use Quick Settings from the lock screen: 1. Wake your iPhone by tapping the screen or pressing the side button. 2. On an iPhone with Face ID, swipe down from the top-right corner of the screen to open the Control Center. On an iPhone with a Home button, swipe up from the bottom of the screen. 3. Adjust settings such as toggling Wi-Fi, enabling Low Power Mode, adjusting brightness, or controlling music playback without unlocking the device. 4. When finished, simply tap the lock screen or press the side button to exit the Control Center.

Use Case: Imagine you're in a meeting and your phone is locked on the table. A notification sounds, and you want to quickly silence the phone without unlocking it. By swiping down to open Quick Settings from the lock screen, you can instantly enable Silent Mode or Do Not Disturb, ensuring minimal disruption while maintaining security.

Tip 13: Use Haptic Touch

Haptic Touch provides tactile feedback when you press and hold certain elements on your iPhone screen, allowing you to quickly access additional options without opening an app fully. It enhances navigation by offering shortcuts, previews, and contextual menus through subtle vibrations.

Watch Tutorial:
Tap here
https://link.content360.io/use_haptic_touch

Figure 13: Scan to Watch

To use Haptic Touch:
1. Press and hold an app icon, notification, or link on your iPhone screen.
2. Wait for a slight vibration indicating the Haptic Touch menu has appeared.
3. Explore the available options or shortcuts that appear in the contextual menu.
4. Tap the desired option to perform an action or preview content.

For example, pressing and holding the Camera app icon lets you quickly take a selfie, record a video, or open the camera directly. This saves time by providing fast access to common functions without navigating through multiple screens.

Everyday Essentials

Tip 14: Use Flashlight

The iPhone flashlight is a quick and convenient tool that uses the device's built-in LED flash to provide bright, steady light. It can be used in dark environments to illuminate your surroundings without needing an external light source.

Watch Tutorial:
Tap here
https://link.content360.io/iphone-flashlight

Figure 14: Scan to Watch

To turn on the flashlight, swipe down from the top-right corner of the screen (on iPhone models with Face ID) or swipe up from the bottom edge of the screen (on models with a Home button) to open the Control Center. Tap the flashlight icon to activate the light. To turn it off, simply tap the icon again or close the Control Center. You can also access the flashlight from the lock screen by pressing and holding the flashlight icon in the bottom-left corner.

Use the flashlight when you need quick illumination, such as finding your way in a dark room, locating dropped items at night, or reading in low light without disturbing others. It's a handy feature that provides immediate light without having to unlock your phone or open an app.

Tip 15: Use Low Power Mode

Low Power Mode is a feature on the iPhone designed to extend battery life by reducing background activity and visual effects when your battery is running low. When enabled, it temporarily limits mail fetch, background app refresh, automatic downloads, and some visual effects until you can fully charge your device.

Watch Tutorial:
Tap here
https://link.content360.io/low_power_mode

Figure 15: Scan to Watch

To activate Low Power Mode: 1. Open the Settings app on your iPhone. 2. Scroll down and tap on Battery. 3. Toggle the switch next to Low Power Mode to turn it on. Alternatively, you can add Low Power Mode to the Control Center for quick access by going to Settings > Control Center > Customize Controls, then tapping the plus icon next to Low Power Mode.

Use Low Power Mode when you know you will be away from a charger for an extended period, such as during travel or long workdays. It helps conserve battery life, ensuring your iPhone remains functional until you can recharge it, without having to manually close apps or reduce usage drastically.

Tip 16: Use Do Not Disturb Mode

Do Not Disturb mode silences calls, alerts, and notifications while your iPhone is locked, allowing you to focus or rest without interruptions. You can customize it to allow calls from certain contacts or to activate automatically at scheduled times.

Watch Tutorial:
Tap here
https://link.content360.io/use_do_not_disturb

Figure 16: Scan to Watch

Steps: 1. Open the Settings app and tap on "Do Not Disturb." 2. Toggle on "Do Not Disturb" to enable it immediately or set a schedule by tapping "Scheduled" and choosing your preferred start and end times. 3. Adjust settings such as allowing calls from favorites or repeated calls to come through. 4. Alternatively, swipe down from the top-right corner of the screen to open Control Center and tap the crescent moon icon to quickly enable or disable Do Not Disturb.

Use Case: When you are in a meeting or sleeping, turning on Do Not Disturb mode prevents distractions from notifications and calls, helping you maintain focus or get uninterrupted rest. You can also schedule it to activate automatically during your usual sleep hours.

Tip 17: Use Night Shift

Night Shift adjusts the colors of your iPhone display to the warmer end of the spectrum after sunset. This reduces blue light exposure, which can interfere with your sleep by affecting your body's natural circadian rhythm.

Watch Tutorial:
Tap here
https://link.content360.io/iphone-night-shift-display

Figure 17: Scan to Watch

To enable Night Shift: 1. Open the Settings app on your iPhone. 2. Tap on Display & Brightness. 3. Select Night Shift. 4. You can schedule it to turn on automatically from sunset to sunrise or set custom hours. 5. Adjust the color temperature slider to make the display warmer or cooler according to your preference. 6. Alternatively, you can enable Night Shift manually from the Control Center by pressing firmly (or long pressing) on the brightness control and tapping the Night Shift button.

Use Night Shift when using your iPhone in the evening or before bedtime to minimize blue light exposure, helping you fall asleep more easily and improving overall sleep quality.

Tip 18: Use The Control Center Timer

The Control Center Timer on an iPhone allows you to quickly set a countdown without opening the Clock app. This feature is convenient for managing short tasks, cooking, or timing breaks, providing easy access right from the home screen or any app.

Watch Tutorial:
Tap here
https://link.content360.io/set_timer

Figure 18: Scan to Watch

To use the Control Center Timer: 1. Swipe down from the top-right corner of the screen (on iPhone models with Face ID) or swipe up from the bottom edge (on models with a Home button) to open the Control Center. 2. Locate the Timer icon, which looks like a clock or hourglass. 3. Press and hold the Timer icon to open the timer interface. 4. Set the desired countdown duration by scrolling through the hour, minute, and second wheels. 5. Tap "Start" to begin the countdown. 6. To cancel or pause the timer, open the Control Center again and tap the Timer icon.

Use Case: If you are boiling eggs and want to ensure they cook for exactly seven minutes, you can quickly start a timer from the Control Center without interrupting what you're doing on your phone. This fast access helps you stay on track without having to navigate away from your current app.

Tip 19: Customize The Control Center

The Control Center on your iPhone provides quick access to commonly used settings and apps, allowing you to adjust brightness, volume, connectivity options, and more without navigating through multiple menus. Customizing the Control Center lets you add, remove, or rearrange controls to suit your personal preferences and streamline your workflow.

Watch Tutorial:
Tap here
https://link.content360.io/customize_control_center

Figure 19: Scan to Watch

Steps:
1. Open the Settings app on your iPhone.
2. Tap on "Control Center."
3. To add controls, scroll down to the "More Controls" section and tap the green plus (+) icon next to the features you want to include.
4. To remove controls, tap the red minus (-) icon next to the items under "Included Controls," then tap "Remove."
5. To rearrange controls, press and hold the three horizontal lines next to an included control and drag it to your preferred position.
6. Exit Settings. Swipe down from the top-right corner of the screen (or swipe up from the bottom on older models) to open your customized Control Center.

Use Case
If you frequently use the flashlight, calculator, or screen recording features, adding these controls to the Control Center allows for instant access without having to open specific apps. This is particularly useful in situations where you need to quickly capture a screenshot or adjust brightness in changing lighting conditions. Customizing the Control Center ensures your most-used tools are always at your fingertips.

Tip 20: Use Control Center For Settings

Control Center provides quick access to essential settings and controls on your iPhone, allowing you to adjust brightness, volume, connectivity options, and more without navigating through multiple menus. It streamlines everyday tasks, saving time and improving usability.

Watch Tutorial:
Tap here
https://link.content360.io/Control_center_settings

Figure 20: Scan to Watch

To use Control Center for settings: 1. On an iPhone with Face ID, swipe down from the top-right corner of the screen. On iPhones with a Home button, swipe up from the bottom edge of the screen. 2. Tap the icons to toggle settings such as Wi-Fi, Bluetooth, Airplane Mode, and Do Not Disturb. 3. Use sliders to adjust brightness and volume. 4. Press and hold certain controls, like the brightness or volume slider, to reveal additional options. 5. Customize which controls appear in Control Center by going to Settings > Control Center > Customize Controls.

When you're in a meeting and need to quickly silence your phone, the Control Center lets you enable Do Not Disturb mode instantly. Similarly, if you want to connect to a Wi-Fi network or turn on Bluetooth for wireless headphones, Control Center provides a fast and convenient way to manage these settings without interrupting your current activity.

Tip 21: Use Screen Time

Screen Time is a feature on the iPhone that helps you monitor and manage how much time you spend on your device and in specific apps. It provides detailed reports and allows you to set limits to promote healthier digital habits.

Watch Tutorial:
Tap here
https://link.content360.io/use_screen_time

Figure 21: Scan to Watch

To use Screen Time:

1. Open the Settings app on your iPhone.
2. Tap on "Screen Time."
3. If it's your first time, tap "Turn On Screen Time" and follow the prompts.
4. Review the daily and weekly activity reports to see your usage patterns.
5. To set app limits, tap "App Limits," then "Add Limit."
6. Choose the app categories or specific apps you want to restrict.
7. Set the desired time limit and tap "Add."
8. Use "Downtime" to schedule periods when only calls and allowed apps are accessible.
9. Enable "Always Allowed" to select apps you want to use during Downtime.
10. For additional control, set a Screen Time passcode to prevent changes.

For example, if you notice spending excessive time on social media, you can set a daily limit of 30 minutes for those apps. When the limit is reached, Screen Time will notify you and block access until the next day, helping you reduce distractions and focus on other tasks.

Tip 22: Use The Wallet App

The Wallet app on your iPhone allows you to store and manage your credit and debit cards, boarding passes, tickets, rewards cards, and more in one convenient place. This eliminates the need to carry physical cards and provides quick access when you need them. You can also use it to make secure payments with Apple Pay, which works anywhere contactless payments are accepted.

Watch Tutorial:
Tap here
https://link.content360.io/wallet_app

Figure 22: Scan to Watch

To add a card or pass to the Wallet app:
1. Open the Wallet app on your iPhone.
2. Tap the plus (+) button in the top right corner.
3. Follow the prompts to add a credit or debit card by either scanning the card or entering the details manually. For tickets or passes, you can add them by scanning a QR code or tapping a link sent to you.
4. Verify your card with your bank or card issuer if required.
5. Once added, you can access your cards and passes anytime by opening the Wallet app or double-clicking the side button (on supported iPhones) when your phone is locked.

Use the Wallet app when you want to make a quick purchase without pulling out your physical wallet, such as paying at a coffee shop with Apple Pay. It's also handy when traveling—store boarding passes and hotel keys in Wallet for easy, paperless access during your trip. This keeps essential items organized and accessible right from your iPhone.

Tip 23: Use Apple Pay

Apple Pay allows you to make secure and contactless payments using your iPhone. It stores your credit, debit, and prepaid cards digitally, enabling quick transactions without the need for physical cards or cash. Apple Pay also works within apps and on websites, streamlining the checkout process while protecting your payment information through encryption and biometric authentication.

Watch Tutorial:
Tap here
https://link.content360.io/use_apple_pay

Figure 23: Scan to Watch

To set up and use Apple Pay: 1. Open the Wallet app on your iPhone. 2. Tap the plus sign (+) to add a new card. 3. Follow the prompts to scan your card or enter details manually. 4. Verify your card with your bank if required. 5. When ready to pay in stores, double-click the side button (or Home button on older models) to bring up Apple Pay. 6. Authenticate using Face ID, Touch ID, or your passcode. 7. Hold the top of your iPhone near the contactless payment terminal until you see a checkmark and a confirmation sound. 8. For in-app or online purchases, select Apple Pay as the payment method and complete authentication.

Use Apple Pay to speed up your checkout experience when shopping in stores, ordering food, or making purchases within apps. It reduces the need to carry physical wallets and enhances security by keeping your card details private during transactions.

Tip 24: Use The Weather App Widgets

The Weather app widgets provide quick access to current weather conditions and forecasts right from your iPhone's Home Screen or Today View. These widgets update automatically, allowing you to stay informed about the weather without opening the app. You can customize the widget size and location to fit your needs, making it easier to plan your day based on real-time weather information.

Watch Tutorial:
Tap here
https://link.content360.io/weather_app

Figure 24: Scan to Watch

To add the Weather app widget to your Home Screen or Today View: 1. Touch and hold an empty area on your Home Screen until the apps start to jiggle. 2. Tap the plus (+) button in the top-left corner. 3. Scroll through the widget options and select the Weather app. 4. Swipe to choose the widget size you prefer (small, medium, or large). 5. Tap "Add Widget" to place it on your Home Screen. 6. Drag the widget to your preferred location and tap "Done" to exit editing mode.

Use the Weather app widgets to quickly check the temperature, precipitation chances, and weather conditions throughout the day without unlocking your phone or opening the Weather app. This is especially useful when you're on the go and need to decide whether to bring an umbrella or dress warmly before heading outside.

Tip 25: Use Focus Modes

Focus Modes help you minimize distractions by customizing notifications and app access based on your current activity or location. By setting up different Focus Modes, you can create tailored environments for work, personal time, sleep, or any situation where you want to limit interruptions and stay more focused.

Watch Tutorial:
Tap here
https://link.content360.io/use_focus_app

Figure 25: Scan to Watch

Steps:
1. Open the Settings app on your iPhone.
2. Tap on "Focus."
3. Choose a preset Focus Mode such as Work, Personal, Sleep, or create a new one by tapping the plus (+) icon.
4. Customize who can notify you by selecting contacts under "Allowed Notifications."
5. Select which apps are allowed to send notifications during this Focus Mode.
6. Optionally, set a schedule or location-based activation for the Focus Mode.
7. Turn on the Focus Mode manually from the Control Center or let it activate automatically based on your settings.

Use Case:
When starting your workday, activate the Work Focus Mode to silence personal app notifications and only allow calls or messages from colleagues. This helps you maintain concentration on tasks without being distracted by social media or unrelated alerts. In the evening, switching to a Personal Focus Mode can mute work-related notifications, enabling you to relax and spend quality time with family or friends.

Tip 26: Use Optimized Battery Charging

Optimized Battery Charging is a feature designed to extend the lifespan of your iPhone's battery by reducing the time it spends fully charged. When enabled, your iPhone learns your daily charging routine and delays charging past 80% until just before you need to use it. This helps minimize battery aging caused by prolonged exposure to high charge levels.

Watch Tutorial:
Tap here
https://link.content360.io/Optimize_battery_charging

Figure 26: Scan to Watch

To enable Optimized Battery Charging:

1. Open the Settings app on your iPhone.
2. Scroll down and tap on Battery.
3. Select Battery Health & Charging.
4. Toggle on the switch for Optimized Battery Charging.

If you typically charge your phone overnight, the iPhone will hold the charge at 80% and complete the remaining 20% shortly before you wake up, ensuring your battery is at full capacity when you start your day.

Use this feature if you charge your iPhone regularly overnight or for extended periods, as it helps maintain battery health over time, reducing the need for premature battery replacements and keeping your device running efficiently longer.

Camera & Photos

Tip 27: Use The Camera Self Timer

The Camera Self Timer on your iPhone allows you to set a delay between pressing the shutter button and when the photo is actually taken. This feature helps reduce camera shake caused by tapping the screen and gives you time to get into position for group shots or selfies without holding the phone.

Watch Tutorial:
Tap here
https://link.content360.io/iphone-camera-self-timer

Figure 27: Scan to Watch

To use the Camera Self Timer: 1. Open the Camera app on your iPhone. 2. Tap the timer icon, which looks like a clock, located at the top of the screen. 3. Choose the timer duration—either 3 seconds or 10 seconds. 4. Frame your shot and press the shutter button. 5. The timer will count down, giving you time to prepare before the photo is captured.

Use the Camera Self Timer when you want to take group photos without leaving anyone out, or when taking selfies with both hands free. It is especially helpful for steady, blur-free pictures when placing the iPhone on a stable surface or tripod.

Tip 28: Use The Camera Burst Mode

Burst Mode allows you to capture a series of photos in quick succession by holding down the shutter button. This feature is especially useful for photographing fast-moving subjects or capturing the perfect moment in a sequence of actions. Instead of taking a single photo, Burst Mode rapidly takes multiple shots, giving you the option to select the best image later.

Watch Tutorial:
Tap here
https://link.content360.io/iphone-camera-burst-mode

Figure 28: Scan to Watch

To use Burst Mode on an iPhone: 1. Open the Camera app. 2. Point the camera at your subject. 3. Press and hold the shutter button (the white circle) instead of tapping it. 4. Keep holding to capture a burst of photos. 5. Release the shutter button to stop capturing. 6. Open the Photos app, find the burst, tap "Select," and choose your favorite shots to keep. The rest can be deleted to save space.

Use Burst Mode when photographing children playing, pets running, or any situation with fast movement where a single shot might miss the desired expression or action. It helps ensure you get a clear, well-timed photo without the need to retake the shot multiple times.

Tip 29: Use The Camera Grid

The Camera Grid on your iPhone helps you align shots using the rule of thirds, a photography principle that divides your frame into nine equal parts with two horizontal and two vertical lines. This grid makes it easier to compose balanced and visually appealing photos by positioning key elements along these lines or their intersections, improving your shot's overall composition and focus.

Watch Tutorial:
Tap here
https://link.content360.io/iPhone-camera-grid

Figure 29: Scan to Watch

To enable and use the Camera Grid: 1. Open the Settings app on your iPhone. 2. Scroll down and tap on "Camera." 3. Toggle on the "Grid" option. 4. Open the Camera app, and you will see the grid lines overlaying the screen. 5. Use these lines to align subjects or horizons, ensuring your photo is straight and well-composed.

For example, when taking a landscape photo, you can align the horizon with one of the horizontal grid lines to keep it straight, while placing a tree or building at one of the vertical intersections to create a more dynamic and balanced image. This technique helps make your photos look more professional and visually engaging without the need for advanced editing.

Tip 30: Use The Camera Level

The Camera Level feature on your iPhone helps you capture perfectly aligned photos by showing a guide that ensures your shot is level. This is especially useful for landscape photography, architectural shots, or any image where straight lines and balanced composition are important.

Watch Tutorial:
Tap here
https://link.content360.io/iphone-camera-level

Figure 30: Scan to Watch

To use the Camera Level: 1. Open the Camera app on your iPhone. 2. Switch to Photo mode and frame your shot. 3. Slowly move your phone until the on-screen level indicator appears, typically as a horizontal bar or crosshair, signaling that your device is perfectly aligned. 4. When the indicator shows your phone is level, tap the shutter button to take the photo.

Use the Camera Level when photographing horizons, buildings, or any scene where a tilted angle would detract from the image quality. This feature ensures your photos look professional and visually pleasing without needing post-editing adjustments.

Tip 31: Use Live Photo

Live Photo captures a few seconds of motion and sound before and after you press the shutter button, bringing your photos to life with movement and audio. This feature allows you to relive moments more vividly than a still image alone.

Watch Tutorial:
Tap here
https://link.content360.io/iphone-live-photos

Figure 31: Scan to Watch

To use Live Photo: 1. Open the Camera app on your iPhone. 2. Ensure the Live Photo icon (a set of concentric circles) at the top of the screen is yellow, indicating it is enabled. If it's white with a slash, tap it to turn Live Photo on. 3. Press the shutter button to take a picture. The iPhone will record 1.5 seconds before and after the shot. 4. To view the Live Photo, open the Photos app, select the image, and press firmly (or press and hold on devices without 3D Touch) to see the animation.

Use Live Photo to capture moments with subtle movements or sounds, such as a child blowing out birthday candles or waves gently crashing on a beach, making your photos more dynamic and memorable.

Tip 32: Use Portrait Mode

Portrait Mode on the iPhone allows you to take photos with a sharp subject and a beautifully blurred background, creating a professional-looking depth effect. This feature uses advanced camera technology to simulate the shallow depth of field typically achieved with DSLR cameras, making your photos stand out with a cinematic quality.

Watch Tutorial:
Tap here
https://link.content360.io/use_portrait_mode

Figure 32: Scan to Watch

To use Portrait Mode: 1. Open the Camera app on your iPhone. 2. Swipe to select "Portrait" from the camera modes at the bottom. 3. Frame your subject within the yellow focus box that appears. 4. Adjust the lighting effect if desired by tapping the lighting icon and selecting from options like Natural Light or Studio Light. 5. Tap the shutter button to capture your photo.

Portrait Mode is ideal for taking striking photos of people, pets, or objects where you want the focus to be on the subject while the background softly fades away. For example, use Portrait Mode when capturing a friend's portrait at a party, emphasizing their expression while minimizing distracting elements behind them.

Tip 33: Use Cinematic Mode

Cinematic Mode on the iPhone allows you to create videos with a shallow depth of field and automatic focus changes, giving your videos a professional, film-like quality. It intelligently tracks subjects, shifting focus smoothly between them to highlight the most important elements in your scene.

Watch Tutorial:
Tap here
https://link.content360.io/cinematic_mode

Figure 33: Scan to Watch

To use Cinematic Mode: 1. Open the Camera app on your iPhone. 2. Swipe to select "Cinematic" mode. 3. Frame your shot and tap the record button to start filming. 4. As you record, the iPhone will automatically adjust focus between subjects in the frame. 5. After recording, you can tap on different subjects in the video to change the focus manually. 6. Edit the focus points and depth effect in the Photos app if desired.

Use Cinematic Mode when capturing moments like interviews, storytelling videos, or family gatherings. It helps draw attention to people or objects by blurring the background and smoothly shifting focus, making your home videos look more polished and engaging.

Tip 34: Use Photo Cutout

Photo Cutout allows you to isolate a subject from an image by removing the background, creating a transparent cutout that you can use in other apps or share with others. This feature leverages on-device intelligence to quickly detect and separate the main subject, making it easy to create stickers, collages, or custom graphics without needing third-party software.

Watch Tutorial:
Tap here
https://link.content360.io/cut_out_mode

Figure 34: Scan to Watch

Steps: 1. Open the Photos app on your iPhone and select the photo containing the subject you want to cut out. 2. Tap and hold on the subject within the image until you see a glowing outline appear around it. 3. Release your finger, and a menu will pop up with options like Copy, Share, and Save. 4. Choose "Copy" to copy the cutout to your clipboard or "Share" to send it directly to another app. 5. You can then paste the cutout into apps like Messages, Notes, or third-party editing apps to use it as a sticker or graphic element.

Use Case: Imagine you took a picture of your pet and want to create a fun sticker to send to friends. Using Photo Cutout, you can isolate your pet from the background in seconds, then paste the cutout into a messaging app or photo editor to create personalized stickers and share them easily. This saves time and effort compared to manually cropping or using complex editing tools.

Tip 35: Use Visual Look Up

Visual Look Up is a feature on the iPhone that helps you learn more about objects, landmarks, plants, animals, and other items in your photos by analyzing the image and providing relevant information. It can identify things such as famous art, flowers, books, or even breeds of dogs, offering instant knowledge without the need to search manually.

Watch Tutorial:
Tap here
https://link.content360.io/Visual_look_up

Figure 35: Scan to Watch

To use Visual Look Up: 1. Open the Photos app and select a photo that contains a recognizable object or scene. 2. Look for the Visual Look Up icon, which appears as a glowing star or info button, usually at the bottom right corner of the photo. 3. Tap the icon to see information related to the subject in the photo. 4. Browse through the details, links, or suggestions provided to learn more.

Visual Look Up is useful when you come across an unfamiliar plant during a hike and want to identify it quickly. Instead of searching the internet, you can simply take a photo and use Visual Look Up to get instant facts, helping you satisfy your curiosity or ensure safety by recognizing potentially harmful species.

Tip 36: Use Live Text In Photos

Live Text allows you to interact with text directly within your photos. You can copy, paste, look up, translate, or share text found in images, making it easier to extract information without typing it manually.

Watch Tutorial:
Tap here
https://link.content360.io/Use_live_text

Figure 36: Scan to Watch

Steps: 1. Open the Photos app and select a photo containing text. 2. Tap and hold on the text within the photo until you see selection handles. 3. Adjust the selection handles to highlight the desired text. 4. Use the options that appear, such as Copy, Look Up, Translate, or Share, to interact with the text. 5. Alternatively, tap the Live Text icon (a small text box) in the bottom right corner of the photo to quickly access text recognition features.

Use Case: Imagine receiving a photo of a restaurant menu or a handwritten note. Instead of typing the details manually, you can use Live Text to quickly copy the phone number or address and paste it into Maps or Messages, saving time and reducing errors.

Tip 37: Use The Photo App

The Photos app on your iPhone is a powerful tool for organizing, editing, and sharing your images and videos. It automatically sorts your media into albums and memories, making it easy to find specific photos. The app also offers a variety of editing tools to enhance your pictures without needing third-party software.

Watch Tutorial:
Tap here
https://link.content360.io/use_photo_app

Figure 37: Scan to Watch

To use the Photos app effectively:
1. Open the Photos app from your home screen.
2. Browse your photos by tapping on the "Photos" tab for a chronological view or the "Albums" tab to see grouped collections.
3. To edit a photo, select it and tap "Edit" in the top right corner. Use the available tools to adjust light, color, crop, and apply filters.
4. Tap "Done" to save your changes.
5. To share a photo, tap the share icon and choose from options like AirDrop, Messages, Mail, or social media apps.

If you want to quickly find a picture from a recent vacation, you can switch to the "Albums" tab and select the automatically created "Places" or "People" albums to locate images by location or faces recognized by your iPhone. This makes it easy to organize and relive your memories without scrolling through hundreds of photos.

Tip 38: Use ProRAW & ProRes

ProRAW and ProRes are advanced photo and video formats available on certain iPhone models, designed to provide higher quality and greater editing flexibility. ProRAW captures more image data than standard photos, preserving details and dynamic range, which is ideal for professional-level photo editing. ProRes is a high-quality video codec that maintains superior color fidelity and detail, making it suitable for video projects that require extensive post-production work.

Watch Tutorial:
Tap here
https://link.content360.io/Use-ProRAW-ProRes

Figure 38: Scan to Watch

To enable and use ProRAW: 1. Open the Settings app and go to Camera. 2. Tap on Formats. 3. Toggle on Apple ProRAW. 4. Open the Camera app and select Photo mode. 5. Tap the RAW button at the top right to enable or disable ProRAW for each shot.

To enable and use ProRes: 1. Open the Settings app and go to Camera. 2. Tap on Formats. 3. Toggle on Apple ProRes. 4. Open the Camera app and select Video mode. 5. Tap the ProRes icon to turn it on or off before recording.

A professional photographer capturing a sunset may use ProRAW to retain maximum detail and color information, allowing for fine adjustments in editing software without degrading image quality. Similarly, a videographer recording a short film could use ProRes to ensure the footage maintains optimal clarity and color depth during post-production, resulting in a polished final product.

Tip 39: Use Photographic Styles

Photographic Styles allow you to customize the look of your photos directly in the Camera app on your iPhone without affecting the overall tone and color of the scene. Unlike filters that apply a uniform effect, Photographic Styles intelligently adjust specific parts of the image, such as skin tones or shadows, to maintain a natural appearance while enhancing the photo according to your preferred style.

Watch Tutorial:
Tap here
https://link.content360.io/Use-Photographic-Styles

Figure 39: Scan to Watch

To use Photographic Styles:

1. Open the Camera app on your iPhone.
2. Swipe up or tap the arrow at the top of the screen to reveal additional camera settings.
3. Tap the Photographic Styles icon (a square with lines) located in the toolbar.
4. Browse through the available styles such as Standard, Rich Contrast, Vibrant, Warm, and Cool.
5. Select the style that best suits your desired look; you can also customize the tone and warmth sliders for a more personalized effect.
6. Compose and capture your photo as usual; the style will be applied in real-time.

Use Photographic Styles when you want to create photos with a consistent aesthetic directly from the Camera app, eliminating the need for post-processing edits. For example, if you prefer warmer tones for portraits, selecting a Warm style will enhance skin tones beautifully while preserving natural details, making your images look polished and professional instantly.

Communication & Apps

Tip 40: Use FaceTime

FaceTime is Apple's built-in video and audio calling service that allows iPhone users to connect with others using an internet connection. It supports both one-on-one and group calls, providing high-quality video and clear audio. FaceTime works seamlessly across iPhone, iPad, iPod touch, and Mac devices, making it easy to stay in touch with friends, family, or colleagues.

Watch Tutorial:
Tap here
https://link.content360.io/iphone-facetime-video-calls

Figure 40: Scan to Watch

To use FaceTime on your iPhone:

1. Open the FaceTime app.
2. Tap the plus (+) button in the top-right corner.
3. Enter the phone number or email address of the person you want to call. You can add multiple contacts for a group call.
4. Choose either the video call button (camera icon) or the audio call button (phone icon) to start the call.
5. During the call, you can use features like mute, switch camera, or add more participants.

FaceTime is ideal for staying connected when in-person meetings are not possible. For example, you can use it to catch up with family members who live far away or hold virtual meetings with coworkers while working remotely. Its easy interface and reliable connection make it a convenient tool for both personal and professional communication.

Tip 41: Use iMessage

iMessage allows you to send texts, photos, videos, and more to other Apple users over Wi-Fi or cellular data without using traditional SMS. Messages sent via iMessage appear in blue bubbles, indicating they are encrypted and sent through Apple's servers, providing a faster and more secure messaging experience compared to standard text messages.

Watch Tutorial:
Tap here
https://link.content360.io/iphone-imessage-features

Figure 41: Scan to Watch

To use iMessage:

1. Open the Messages app on your iPhone.
2. Tap the compose button (a pencil inside a square) to start a new message.
3. Enter the recipient's phone number or Apple ID email address. If the contact uses iMessage, their name will appear in blue.
4. Type your message in the text box.
5. Tap the send button (blue arrow) to deliver the message.
6. You can also tap the camera icon to attach photos or videos, or tap the App Store icon to add stickers, GIFs, or other interactive content.

Use iMessage to stay connected with friends and family who use Apple devices. Since iMessage supports features like read receipts, typing indicators, and high-quality media sharing, it enhances communication by making conversations more interactive and engaging without incurring SMS fees. For example, you can quickly share a group photo from a recent trip with multiple friends, and everyone will receive it instantly in high resolution.

Tip 42: Use AirDrop

AirDrop allows you to quickly and securely share photos, documents, and other files between Apple devices without needing an internet connection. It uses Bluetooth to create a peer-to-peer Wi-Fi network between devices, ensuring fast and private transfers.

Watch Tutorial:
Tap here
https://link.content360.io/iphone-airdrop-file-sharing

Figure 42: Scan to Watch

Steps: 1. Open the file or item you want to share on your iPhone. 2. Tap the Share icon (a square with an upward arrow). 3. Select AirDrop from the sharing options. 4. Choose the nearby Apple device you want to send the file to. 5. The recipient will receive a prompt to accept or decline the transfer. 6. Once accepted, the file will be transferred instantly.

Use Case: When you want to quickly share a photo from your recent vacation with a friend standing next to you, AirDrop lets you send the image without fumbling with email or messaging apps. This is especially useful in environments without reliable internet or when you want to maintain privacy during file transfers.

Tip 43: Use The Note App

The Notes app on the iPhone is a versatile tool for quickly capturing ideas, making lists, and organizing information in one convenient place. It allows you to create text notes, checklists, sketches, and even scan documents, helping you stay organized and productive throughout your day.

Watch Tutorial:
Tap here
https://link.content360.io/iphone-notes-organization

Figure 43: Scan to Watch

To use the Notes app effectively: 1. Open the Notes app from your home screen. 2. Tap the compose button (a square with a pencil) to create a new note. 3. Type your text, or use the formatting options to add checklists, photos, or sketches by tapping the plus (+) icon. 4. To organize notes, create folders by tapping the back arrow, then New Folder, and assign notes to these folders. 5. Use the search bar at the top to quickly find specific notes. 6. To secure sensitive information, lock a note by tapping the share icon and selecting Lock Note. 7. Sync your notes across devices using iCloud by enabling Notes in Settings > [your name] > iCloud.

Use the Notes app to keep track of your grocery list on the go. Instead of carrying a paper list, create a checklist note that you can update anytime. Check off items as you shop, ensuring you don't forget anything. This also allows you to share the list with family members for collaborative shopping.

Tip 44: Use The Podcast App

The Podcast app on your iPhone allows you to discover, subscribe to, and listen to a wide variety of audio shows, from news and storytelling to educational content and entertainment. It organizes your favorite podcasts in one place, making it easy to keep up with new episodes and listen offline.

Watch Tutorial:
Tap here
https://link.content360.io/iphone-podcasts-listening

Figure 44: Scan to Watch

To use the Podcast app: 1. Open the Podcast app on your iPhone. 2. Browse featured shows or tap the Search tab to find specific podcasts by name or topic. 3. When you find a podcast you like, tap Subscribe to add it to your library. 4. New episodes will automatically appear in your Library tab. 5. Tap an episode to start playing it, or download it for offline listening by tapping the download icon. 6. Use the playback controls to pause, skip, or adjust playback speed.

If you commute daily, subscribing to your favorite podcasts can turn travel time into an opportunity to learn, stay informed, or be entertained, all without needing an internet connection once episodes are downloaded.

Tip 45: Use The Apple TV App

The Apple TV app consolidates your favorite movies and TV shows from various streaming services into one convenient location, making it easier to browse, watch, and manage your entertainment. It also offers personalized recommendations based on your viewing habits and allows you to rent or purchase movies and TV shows directly within the app.

Watch Tutorial:
Tap here
https://link.content360.io/iphone-apple-tv-streaming

Figure 45: Scan to Watch

Steps: 1. Open the Apple TV app on your iPhone. 2. Browse through the different tabs such as Watch Now, Library, and Store to find content. 3. Use the Search tab to look for specific movies, shows, or channels. 4. Tap on a title to see details, watch trailers, or start playback. 5. If the content requires purchase or rental, follow the prompts to complete the transaction. 6. Add shows or movies to your Up Next queue for easy access later. 7. Link your streaming service accounts within the app to access more content in one place.

Use Case: If you subscribe to multiple streaming services like Apple TV+, Hulu, and CBS All Access, the Apple TV app lets you seamlessly browse and watch shows from all these platforms without switching apps. This saves time and helps you discover new content based on your preferences, all while keeping your watchlist organized in a single app.

Tip 46: Use SharePlay

SharePlay allows you to enjoy synchronized content like movies, music, or workouts with friends and family during a FaceTime call. It lets everyone watch or listen together in real-time, even when apart, creating a shared experience.

Watch Tutorial:
Tap here
https://link.content360.io/Use-SharePlay

Figure 46: Scan to Watch

To use SharePlay: 1. Start a FaceTime call with one or more contacts. 2. Open a supported app, such as Apple TV, Apple Music, or a fitness app. 3. Begin playing the content you want to share. 4. When prompted, tap "SharePlay" to start the shared session. 5. Everyone on the call can play, pause, or control the content, and chat at the same time.

Use SharePlay to watch a new movie with friends who live far away, making it feel like you're in the same room. It's also great for listening to music playlists together or following a workout routine simultaneously, keeping motivation high and connection strong.

Tip 47: Use NameDrop

NameDrop allows you to quickly and securely share your contact information, photos, or files with another iPhone by simply bringing both devices close together. This feature uses a combination of Bluetooth and Ultra Wideband technology to establish a direct connection, making the exchange seamless and private.

Watch Tutorial:
Tap here
https://link.content360.io/Use-NameDrop

Figure 47: Scan to Watch

To use NameDrop: 1. Ensure both iPhones have NameDrop enabled and are running iOS 17 or later. 2. Bring the two iPhones close to each other, near the top where the TrueDepth camera is located. 3. A NameDrop prompt will appear on both screens, allowing you to choose what information or files you want to share. 4. Tap the desired content to send, and the recipient can accept or decline the transfer. 5. Once accepted, the information is added directly to the recipient's device without needing to open any apps.

Use NameDrop when meeting new contacts at events or social gatherings to instantly exchange phone numbers, email addresses, or social profiles without manually typing or sending messages. It's also handy for sharing photos or documents with friends nearby without the hassle of AirDrop menus.

Tip 48: Use Check In Safety Feature

The Check In safety feature on your iPhone allows you to automatically notify a chosen contact when you arrive safely at a destination. This feature is particularly useful when traveling alone or meeting someone new, providing peace of mind to both you and your trusted contacts by sharing your location and status updates.

Watch Tutorial:
Tap here
https://link.content360.io/Use-Check-In-Safety-Feature

Figure 48: Scan to Watch

To use the Check In feature: 1. Open the Messages app and select the conversation with the person you want to notify. 2. Tap the contact's name or profile picture at the top of the screen. 3. Choose "Check In" from the options menu. 4. Set your destination and estimated arrival time. 5. Tap "Start Check In." Your contact will be notified and can follow your progress. 6. If you don't arrive on time or turn off the feature, your location and a message will be automatically shared with the contact.

Use Check In when you're heading to a new place or meeting someone for the first time. For example, if you're traveling to a remote hiking trail, enabling Check In lets a family member know you arrived safely or alerts them if something unexpected happens, ensuring your safety is monitored without constant communication.

Tip 49: Use Live Voicemail

Live Voicemail allows you to see a real-time transcription of a voicemail as the caller leaves it. This feature helps you decide whether to answer the call immediately or let it go to voicemail without having to listen to the message afterward.

Watch Tutorial:
Tap here
https://link.content360.io/Use-Live-Voicemail

Figure 49: Scan to Watch

To use Live Voicemail:
1. Open the Phone app and go to the Voicemail tab.
2. Enable Live Voicemail in your settings if prompted.
3. When someone calls and you don't answer, the caller can leave a voicemail.
4. As the caller speaks, watch the real-time transcript appear on your screen.
5. Choose to answer the call if the message is urgent, or let it continue to voicemail.

Live Voicemail is especially helpful when you're in a meeting or a noisy environment and can't take a call but want to quickly assess if the message requires immediate attention. It lets you stay informed without interrupting your current activity.

Tip 50: Use Contact Posters

Contact Posters allow you to personalize how your contact information appears when you call someone or when they call you. By customizing your Contact Poster, you can add a photo, Memoji, or a unique design along with your name, making your calls more recognizable and visually appealing.

Watch Tutorial:
Tap here
https://link.content360.io/Use-Contact-Posters

Figure 50: Scan to Watch

Steps
1. Open the Phone app and tap your contact card at the top of the Favorites or Recents tab.
2. Tap “Edit” under your name and photo area.
3. Choose “Edit Poster” or “Create New Poster.”
4. Select a photo, Memoji, or background design to use.
5. Customize your name’s font and color to match your style.
6. Tap “Done” to save your Contact Poster.

Use Case
When you call friends or family, your personalized Contact Poster appears on their screen, helping them instantly recognize it’s you. This is especially useful if you share a common name or phone number with others, or if you want to add a personal touch to your calls. It also enhances your own Contacts app experience by making your profile stand out.

Tip 51: Use Stickers From Photos

You can create personalized stickers on your iPhone by using photos from your library. This feature allows you to extract subjects from images and use them as stickers in Messages and other apps, adding a fun and creative way to communicate.

Watch Tutorial:
Tap here
https://link.content360.io/Use-Stickers-From-Photos

Figure 51: Scan to Watch

Steps: 1. Open the Photos app and select the photo containing the subject you want to turn into a sticker. 2. Press and hold on the subject in the photo until a glowing outline appears around it. 3. Drag the extracted subject into the Messages app or any other app that supports stickers. 4. Drop the sticker into the conversation or app interface to send or use it.

Use Case: Imagine you have a picture of your pet with a cute expression. By extracting your pet's image as a sticker, you can quickly send it in a text message to friends or family, making your conversations more lively and personalized without needing to cut and edit images manually.

Productivity & Efficiency

Tip 52: Use The Measure App

The Measure app on your iPhone uses augmented reality (AR) to turn your device into a digital tape measure. It allows you to quickly measure objects and spaces without the need for a physical measuring tool. The app uses the camera and sensors to detect surfaces and calculate distances accurately.

Watch Tutorial:
Tap here
https://link.content360.io/Use-The-Measure-App

Figure 52: Scan to Watch

To use the Measure app: 1. Open the Measure app on your iPhone. 2. Move your iPhone around to let the app detect the surface you want to measure. 3. Position the dot or circle over the starting point of the object or space. 4. Tap the plus (+) button to set the starting point. 5. Slowly move your iPhone to the end point of the area or object. 6. Tap the plus (+) button again to complete the measurement. 7. The app will display the distance between the two points in either inches or centimeters. 8. You can tap the shutter button to capture a photo with the measurement overlaid for future reference.

Use the Measure app when you need a quick measurement on the go, such as checking if furniture will fit in a space, measuring the size of a package before shipping, or estimating dimensions for DIY projects without carrying a tape measure.

Tip 53: Use The Files App

The Files app on your iPhone provides a centralized location to manage, organize, and access all your documents, images, and other files stored locally or in cloud services like iCloud Drive, Google Drive, and Dropbox. It simplifies file management by allowing you to browse, search, and organize files across different storage locations without switching between multiple apps.

Watch Tutorial:
Tap here
https://link.content360.io/Use-The-Files-App

Figure 53: Scan to Watch

Steps: 1. Open the Files app on your iPhone. 2. Tap Browse at the bottom to view different locations such as iCloud Drive, On My iPhone, or connected third-party cloud services. 3. Navigate through folders by tapping on them. 4. To organize files, tap Select, choose the files or folders, then use the folder icon to move, or the trash icon to delete. 5. Use the search bar at the top to quickly find specific files. 6. To create a new folder, tap the folder icon with a plus sign in the upper corner. 7. Tap and hold a file to access options like Rename, Share, or Tag.

Use Case: If you frequently receive documents via email that you want to save for later, the Files app allows you to move those attachments directly into organized folders within iCloud Drive or another cloud service. This makes it easier to locate important documents when you need them, whether for work, school, or personal use, without cluttering your email inbox or needing to open multiple apps.

Tip 54: Use The Shortcuts Widget

The Shortcuts widget on your iPhone provides quick access to your favorite or most-used shortcuts directly from the Home Screen or Today View. This allows you to automate tasks with a single tap, saving time and streamlining your daily routines without needing to open the Shortcuts app each time.

Watch Tutorial:
Tap here
https://link.content360.io/Use-The-Shortcuts-Widget

Figure 54: Scan to Watch

To add and use the Shortcuts widget: 1. On your Home Screen or Today View, press and hold an empty area until the apps jiggle. 2. Tap the plus (+) button in the upper-left corner to open the widget gallery. 3. Scroll or search for "Shortcuts" in the widget list. 4. Choose the widget size you prefer, then tap "Add Widget." 5. Once added, tap the widget to open and run your shortcuts immediately. 6. To customize which shortcuts appear, open the Shortcuts app, arrange or create shortcuts you want quick access to.

For example, if you frequently run a shortcut that sends your ETA to a family member, having the Shortcuts widget on your Home Screen allows you to send your location with one tap, rather than opening the app and navigating through menus. This enhances convenience and speeds up everyday tasks.

Tip 55: Use Car Play

Car Play allows you to connect your iPhone to your car's infotainment system, providing a safer and more convenient way to use your phone while driving. It mirrors select apps from your iPhone on the car's display, enabling you to access navigation, make calls, send messages, and listen to music using voice commands or touch controls.

Watch Tutorial:
Tap here
https://link.content360.io/Use-Car-Play

Figure 55: Scan to Watch

Steps: 1. Ensure your car supports Car Play and that your iPhone is running the latest iOS version. 2. Connect your iPhone to the car using a Lightning cable or wirelessly if your vehicle supports wireless Car Play. 3. On your car's display, select the Car Play icon. 4. Use Siri voice commands, the touchscreen, or car controls to operate compatible apps.

Use Case: While driving to a new location, you can use Car Play to access Apple Maps for turn-by-turn directions without handling your phone. This keeps your focus on the road while still receiving timely navigation prompts and traffic updates.

Tip 56: Use Screen Mirroring

Screen Mirroring allows you to display your iPhone's screen on a compatible device, such as an Apple TV or smart TV. This feature is useful for sharing photos, videos, presentations, or apps on a larger screen for better viewing or collaboration.

Watch Tutorial:
Tap here
https://link.content360.io/iphone-screen-mirroring

Figure 56: Scan to Watch

To use Screen Mirroring:
1. Ensure your iPhone and the target device are connected to the same Wi-Fi network.
2. Open the Control Center by swiping down from the top-right corner of the screen (or up from the bottom on older iPhones).
3. Tap the "Screen Mirroring" icon, which looks like two overlapping rectangles.
4. Select the device you want to mirror your screen to from the list that appears.
5. If prompted, enter the code displayed on the target device to establish the connection.
6. To stop mirroring, open Control Center again and tap "Stop Mirroring."

Use Screen Mirroring when you want to share content during a meeting or family gathering without needing to pass your phone around. It's also helpful for viewing photos or videos on a larger display, making it easier for multiple people to enjoy the content simultaneously.

Tip 57: Use Screen Recording

Screen Recording allows you to capture exactly what happens on your iPhone screen, including any taps, swipes, and app activity. This feature is useful for creating tutorials, saving gameplay, or sharing issues with technical support without needing additional software.

Watch Tutorial:
Tap here
https://link.content360.io/iPhone-Screen-Recording

Figure 57: Scan to Watch

To use Screen Recording on your iPhone: 1. Open the Settings app and go to Control Center. 2. Tap Customize Controls, then add Screen Recording by tapping the green plus button next to it. 3. Swipe down from the top-right corner of the screen (or swipe up from the bottom on older models) to open Control Center. 4. Tap the Screen Recording button (a circle within a circle). After a three-second countdown, recording starts. 5. To stop recording, open Control Center again and tap the Screen Recording button, or tap the red status bar at the top and select Stop. 6. The recording will be saved automatically to your Photos app.

Use Case: If you want to show a friend how to navigate a tricky app or report a bug, screen recording lets you capture the entire process in real time. Instead of describing steps or sending multiple screenshots, you can share a video that clearly demonstrates the issue or instructions.

Tip 58: Use iCloud

iCloud is Apple's cloud storage service that allows you to securely store your photos, documents, app data, and more, keeping them up to date across all your Apple devices. By enabling iCloud, you can easily access your important files from anywhere, ensure your data is backed up, and seamlessly sync information like contacts, calendars, and notes.

Watch Tutorial:
Tap here
https://link.content360.io/iphone-icloud-backup-sync

Figure 58: Scan to Watch

Steps: 1. Open the Settings app on your iPhone. 2. Tap your name at the top to access Apple ID settings. 3. Select "iCloud." 4. Review the list of apps and toggle on the items you want to sync, such as Photos, Contacts, and Notes. 5. To manage storage, tap "Manage Storage" to upgrade your plan or delete unnecessary backups. 6. For device backups, scroll down and tap "iCloud Backup," then enable it and tap "Back Up Now" to create a manual backup.

Use Case: If you take photos on your iPhone and want to access them on your iPad or Mac, enabling iCloud Photos automatically uploads and syncs your pictures across devices. This way, when you capture a memorable moment, it's instantly available on all your Apple devices without needing to transfer files manually.

Tip 59: Use Family Sharing

Family Sharing allows you to share purchases, subscriptions, photos, calendars, and more with up to six family members without sharing accounts. It helps keep everyone connected while maintaining individual privacy and payment methods under one family organizer.

Watch Tutorial:
Tap here
https://link.content360.io/iphone-family-sharing-setup

Figure 59: Scan to Watch

To set up Family Sharing on your iPhone: 1. Open the Settings app and tap your name at the top. 2. Select "Family Sharing" and tap "Set Up Your Family." 3. Follow the on-screen prompts to invite family members by entering their Apple ID emails. 4. Choose the features you want to share, such as Apple Music, iCloud storage, or location sharing. 5. Family members will receive an invitation to join and can accept it on their devices.

Use Family Sharing to share a single Apple Music subscription among family members, allowing everyone to enjoy the service without paying separately. It also simplifies sharing purchased apps and books, reducing individual expenses and keeping family schedules synced with shared calendars.

Tip 60: Use Drag And Drop Between Apps

Drag and drop on the iPhone allows you to move text, images, and files seamlessly between apps without the need to copy and paste. This feature enhances multitasking by enabling you to drag content from one app and drop it directly into another, speeding up your workflow and reducing interruptions.

Watch Tutorial:
Tap here
https://link.content360.io/Use-Drag-And-Drop-Between-Apps

Figure 60: Scan to Watch

Steps: 1. Open the app containing the content you want to move. 2. Touch and hold the item (such as text, a photo, or a file) until it lifts slightly. 3. While holding the item, use another finger to press the Home button or swipe up to open the Dock. 4. Without releasing the dragged item, tap the app you want to drop the content into. 5. Position the item where you want it in the destination app and release your finger to drop it.

Use Case: Imagine you're browsing photos in the Photos app and want to include an image in an email. Instead of saving and attaching the file, you can drag the photo directly from Photos into the body of your Mail message. This saves time and makes transferring content between apps quick and intuitive.

Tip 61: Use Text Replacement Shortcuts

Text Replacement Shortcuts allow you to create custom abbreviations that automatically expand into longer phrases, words, or sentences when typed. This feature saves time and effort by reducing repetitive typing, especially for frequently used expressions like email addresses, common responses, or technical terms.

Watch Tutorial:
Tap here
https://link.content360.io/Use-Text-Replacement-Shortcuts

Figure 61: Scan to Watch

To set up a Text Replacement Shortcut: 1. Open the Settings app on your iPhone. 2. Tap on "General," then select "Keyboard." 3. Choose "Text Replacement." 4. Tap the "+" icon in the upper-right corner. 5. In the "Phrase" field, enter the full text you want to appear. 6. In the "Shortcut" field, type the abbreviation or shortcut you want to use. 7. Tap "Save" to add the shortcut.

For example, if you frequently type your email address, you can create a shortcut like "eml" that expands to your full email. When composing a message or filling out a form, typing "eml" will automatically replace the abbreviation with your email address, speeding up communication and reducing errors.

Tip 62: Use Scan Documents With Notes

Scan Documents With Notes lets you quickly digitize physical documents using your iPhone's camera and store them directly within the Notes app. This feature automatically detects the edges of the document, enhances the image for readability, and allows you to organize, annotate, or share the scanned files without needing a separate app.

Watch Tutorial:
Tap here
https://link.content360.io/Use-Scan-Documents-With-Notes

Figure 62: Scan to Watch

Steps: 1. Open the Notes app on your iPhone and create a new note or open an existing one. 2. Tap the camera icon located above the keyboard. 3. Select "Scan Documents" from the options. 4. Position your iPhone camera over the document; the app will automatically detect the edges and capture the scan. 5. Adjust the corners if necessary, then tap "Keep Scan." 6. Continue scanning additional pages if needed, or tap "Save" when finished. 7. Your scanned document will appear within the note, where you can add text, drawings, or share it via email or messaging apps.

Use Case: You receive a paper receipt at a store and want to save it for expense tracking. Instead of carrying the physical copy, you open Notes, scan the receipt using the Scan Documents feature, and store it alongside your budget notes. This makes it easy to access, reference, and share your receipts digitally whenever required.

Tip 63: Use Handoff Between Devices

Handoff allows you to start a task on one Apple device and seamlessly continue it on another nearby device signed in with the same Apple ID. This feature works with apps like Mail, Safari, Notes, Maps, Messages, Reminders, Calendar, Contacts, and some third-party apps that support Handoff. It enhances productivity by eliminating the need to manually transfer work between devices.

Watch Tutorial:
Tap here
https://link.content360.io/Use-Handoff-Between-Devices

Figure 63: Scan to Watch

To use Handoff between your iPhone and another Apple device:

1. Ensure both devices are signed in to the same Apple ID and connected to Wi-Fi.
2. Turn on Bluetooth and Wi-Fi on both devices.
3. On your iPhone, go to Settings > General > AirPlay & Handoff, then toggle on Handoff.
4. On your other device (such as a Mac or iPad), enable Handoff in the corresponding settings (e.g., System Preferences > General on Mac).
5. Start an activity on one device, like composing an email or browsing a webpage.
6. On the second device, look for the Handoff icon: on a Mac, it appears on the Dock; on an iPad, it shows on the lower-left corner of the Lock screen; on an iPhone, it appears on the app switcher.
7. Tap or click the icon to pick up where you left off.

For example, if you start writing an email on your iPhone but need to finish it on your Mac, open the Mail app on the Mac and click the Handoff icon in the Dock. This instantly opens the draft, allowing you to continue without interruption. This feature is especially helpful when switching from mobile to a desktop environment for more comfortable typing or multitasking.

Privacy & Security

Tip 64: Use Emergency SOS

Emergency SOS allows you to quickly call for help and alert your emergency contacts if you find yourself in a dangerous situation. When activated, your iPhone will automatically call emergency services and send your current location to your emergency contacts.

Watch Tutorial:
Tap here
https://link.content360.io/Use-Emergency-SOS

Figure 64: Scan to Watch

To use Emergency SOS on an iPhone with Face ID, press and hold the side button and either volume button simultaneously until the Emergency SOS slider appears. Drag the slider to call emergency services immediately. Alternatively, continue holding the buttons to trigger an automatic call without sliding. On iPhones with a Home button, rapidly press the side or top button five times to bring up the Emergency SOS slider. You can also set up your emergency contacts in the Health app under Medical ID, so they are notified during an emergency call.

If you are in an accident or feel threatened, you can quickly summon help without needing to unlock your phone or dial a number manually. Emergency SOS ensures that emergency responders and your trusted contacts are alerted promptly, potentially saving valuable time in critical situations.

Tip 65: Use The Find My App

The Find My app on your iPhone helps you locate your Apple devices, share your location with friends and family, and even find lost items with AirTags. It combines the features of Find My iPhone and Find My Friends into a single app, making it easier to track devices and stay connected.

Watch Tutorial:
Tap here
https://link.content360.io/iphone-find-my-device-tracking

Figure 65: Scan to Watch

To use the Find My app:

1. Open the Find My app on your iPhone.
2. Tap the Devices tab to see a list of your Apple devices linked to your Apple ID.
3. Select the device you want to locate. The app will show its current location on a map if it is online.
4. Use the available options such as Play Sound to help locate a nearby device, Directions to get navigation to the device, or Mark as Lost to lock and track a missing device.
5. To share your location with friends or family, tap the People tab, then select Share My Location or add a contact to share with.
6. For AirTags or other compatible accessories, tap the Items tab to view their locations or make them play a sound.

Use Case: If you misplace your iPhone at a coffee shop, open the Find My app on another Apple device or via iCloud.com. Select your iPhone from the Devices list and choose Play Sound to help you locate it nearby. If it's not within earshot, use Directions to navigate to its last known location or mark it as lost to secure your data and receive notifications if someone finds it.

Tip 66: Use Lockdown Mode

Lockdown Mode is a security feature on the iPhone designed to protect users from highly targeted cyberattacks. When enabled, it restricts certain device functionalities and limits interactions that could be exploited by sophisticated hacking attempts, providing an extra layer of defense for users who may be at risk.

Watch Tutorial:
Tap here
https://link.content360.io/Use-Lockdown-Mode

Figure 66: Scan to Watch

To enable Lockdown Mode: 1. Open the Settings app on your iPhone. 2. Scroll down and tap on Privacy & Security. 3. Scroll to the bottom and select Lockdown Mode. 4. Tap Turn On Lockdown Mode. 5. Read the information about the feature and tap Turn On & Restart. 6. Your iPhone will restart and Lockdown Mode will be activated.

Use Lockdown Mode if you believe you are a potential target of advanced cyber threats, such as journalists, activists, or government officials. It is especially useful when traveling to high-risk areas or handling sensitive information, as it minimizes vulnerabilities by limiting device features and data sharing.

Tip 67: Use Safety Check

Safety Check is a feature on the iPhone designed to help users quickly review and manage the access they have granted to others, such as location, privacy permissions, and shared information. It is particularly useful if you feel your personal information is at risk or if you want to reset access granted to someone else for safety reasons. By using Safety Check, you can immediately stop sharing your location, reset privacy permissions, and manage who has access to your passwords and accounts.

Watch Tutorial:
Tap here
https://link.content360.io/Use-Safety-Check

Figure 67: Scan to Watch

Steps: 1. Open the Settings app on your iPhone. 2. Scroll down and tap on "Privacy & Security." 3. Tap on "Safety Check." 4. Choose "Manage Emergency Access" or "Review Sharing and Access." 5. Follow the prompts to stop sharing your location, reset access permissions, or remove people from your shared devices and accounts. 6. Confirm your choices to apply the changes.

Use Case: If you are in a situation where you no longer want a former partner or someone you no longer trust to have access to your location or personal data, using Safety Check allows you to instantly revoke this access without needing to change passwords or manually adjust multiple settings. This tool provides peace of mind by quickly securing your privacy with just a few taps.

Tip 68: Use App Privacy Report

App Privacy Report provides a detailed overview of how apps use the permissions you have granted, showing which apps have accessed sensitive information like your location, camera, microphone, and contacts. This feature helps you monitor app activity and maintain control over your personal data by revealing any unusual or unexpected app behavior.

Watch Tutorial:
Tap here
https://link.content360.io/Use-App-Privacy-Report

Figure 68: Scan to Watch

To use App Privacy Report:

1. Open the Settings app on your iPhone.
2. Scroll down and tap on "Privacy."
3. Select "App Privacy Report."
4. If this is your first time, tap "Turn On App Privacy Report" to start collecting data.
5. After using your apps for a while, return to the App Privacy Report screen to view detailed logs of app activity, including access to sensors, network activity, and website visits.
6. Review the information to identify any apps that may be accessing data more frequently or unexpectedly.

Use App Privacy Report to stay informed about how your apps interact with your personal information. For example, if you notice an app frequently accessing your location in the background without a clear reason, you can choose to revoke its location permission or uninstall the app altogether, enhancing your privacy and security.

Tip 69: Use Mail Privacy Protection

Mail Privacy Protection helps prevent senders from knowing when you open an email and masks your IP address, making it harder for them to track your location and online activity. This feature enhances your privacy by blocking tracking pixels and loading remote content privately.

Watch Tutorial:
Tap here
https://link.content360.io/Use-Mail-Privacy-Protection

Figure 69: Scan to Watch

To enable Mail Privacy Protection: 1. Open the Settings app on your iPhone. 2. Scroll down and tap on Mail. 3. Tap on Privacy Protection. 4. Toggle on "Protect Mail Activity."

Use Case: When you receive newsletters or promotional emails, senders often include tracking pixels to see if and when you open their messages. By enabling Mail Privacy Protection, you prevent these senders from gathering data about your email activity, helping to keep your personal information and browsing habits more secure.

Tip 70: Use Hide My Email

Hide My Email lets you create unique, random email addresses that forward messages to your personal inbox. This feature helps protect your real email address, reducing spam and enhancing privacy when signing up for websites or newsletters.

Watch Tutorial:
Tap here
https://link.content360.io/Use-Hide-My-Email

Figure 70: Scan to Watch

Steps: 1. Open Settings on your iPhone. 2. Tap your Apple ID at the top, then select "iCloud." 3. Choose "Hide My Email." 4. Tap "Create new address" to generate a random email. 5. Label the address for easy identification and tap "Continue." 6. Use this new address when signing up or providing your email online. 7. Manage or deactivate these addresses anytime from the same "Hide My Email" menu.

Use Case: When registering for a new app or newsletter, use Hide My Email to avoid sharing your personal email address. This prevents unwanted emails from cluttering your inbox and helps keep your online identity secure. If spam begins to arrive at the random address, you can simply deactivate it without affecting your primary email.

Tip 71: Use Private Browsing In Safari

Private Browsing in Safari allows you to browse the internet without saving your browsing history, cookies, or other website data. This mode helps protect your privacy, especially when using shared devices or visiting sensitive websites. When Private Browsing is enabled, Safari won't remember the pages you visit, your search history, or autofill information.

Watch Tutorial:
Tap here
https://link.content360.io/Use-Private-Browsing-In-Safari

Figure 71: Scan to Watch

To use Private Browsing in Safari on your iPhone:

1. Open the Safari app.
2. Tap the tabs button, which looks like two overlapping squares, located at the bottom-right corner of the screen.
3. Tap "Private" to switch to Private Browsing mode.
4. Tap the plus (+) icon to open a new private tab.
5. When you finish, you can exit Private Browsing by tapping the tabs button again, selecting "Private" to deselect it, and returning to regular browsing.

Use Private Browsing when you want to keep your online activity confidential, such as when shopping for surprise gifts, researching sensitive topics, or using a device that others also access. This ensures your browsing details remain private and are not saved on your iPhone.

Tip 72: Use Passkeys Instead Of Passwords

Passkeys are a more secure and convenient way to sign in to websites and apps without using traditional passwords. They use cryptographic keys stored on your iPhone, which cannot be stolen or phished like regular passwords. When you log in, your device verifies your identity using Face ID, Touch ID, or your device passcode, providing a seamless and safer authentication method.

Watch Tutorial:
Tap here
https://link.content360.io/Use-Passkeys-Instead-Of-Passwords

Figure 72: Scan to Watch

To use passkeys instead of passwords:

1. Ensure your iPhone is updated to the latest version of iOS.
2. When signing up or logging into a supported website or app, choose the option to use a passkey if available.
3. Follow the on-screen instructions to create or use a passkey, which typically involves authenticating with Face ID, Touch ID, or your passcode.
4. Your passkey will be securely stored in your iCloud Keychain and synced across your Apple devices.
5. For subsequent logins, simply authenticate with Face ID, Touch ID, or your passcode to access your account without entering a password.

Use passkeys when accessing sensitive accounts like banking, email, or social media to enhance security and reduce the risk of password theft. This method also simplifies the login process, eliminating the need to remember or manage complex passwords.

Tip 73: Use Two-Factor Authentication

Two-Factor Authentication (2FA) adds an extra layer of security to your Apple ID and other accounts by requiring not only your password but also a verification code sent to a trusted device. This helps protect your personal information even if someone else knows your password.

Watch Tutorial:
Tap here
https://youtu.be/8NrfcyA3Bf8?si=dLIPA2-KrkU9YJ26

Figure 73: Scan to Watch

To enable Two-Factor Authentication on your iPhone: 1. Open the Settings app. 2. Tap on your name at the top to access Apple ID settings. 3. Select "Password & Security." 4. Tap "Turn On Two-Factor Authentication." 5. Follow the on-screen instructions to verify your trusted phone number and complete the setup.

Using Two-Factor Authentication is especially useful when accessing sensitive apps or services, such as online banking or email, ensuring that even if your password is compromised, unauthorized users cannot log in without the verification code sent to your trusted device.

Tip 74: Use Crash Detection

Crash Detection is a safety feature on the iPhone designed to identify severe car accidents and automatically connect you to emergency services if you are unresponsive. It uses sensors like the accelerometer and gyroscope to detect sudden impacts, helping to ensure timely assistance during critical situations.

Watch Tutorial:
Tap here
https://link.content360.io/Use-Crash-Detection

Figure 74: Scan to Watch

To use Crash Detection:
1. Ensure your iPhone is updated to the latest iOS version, as this feature is available on iPhone 14 and later models.
2. Go to Settings > Emergency SOS.
3. Confirm that Crash Detection is enabled. It is typically turned on by default.
4. In the event of a severe crash, your iPhone will display an alert and sound a loud alarm.
5. If you are unable to respond within about 10 seconds, the iPhone will automatically call emergency services and share your location.

Crash Detection is especially useful for individuals who drive frequently or travel alone, as it provides an added layer of security by ensuring emergency help can be summoned quickly even if the user is incapacitated or unable to make a call.

Tip 75: Use Stolen Device Protection

Use Stolen Device Protection to secure your iPhone in case it is lost or stolen. This feature helps you locate, lock, or erase your device remotely, preventing unauthorized access to your personal information. It works through the Find My app and requires that Find My iPhone is enabled beforehand.

Watch Tutorial:
Tap here
https://link.content360.io/Use-Stolen-Device-Protection

Figure 75: Scan to Watch

To enable Stolen Device Protection: 1. Open the Settings app on your iPhone. 2. Tap your name at the top to access Apple ID settings. 3. Select “Find My.” 4. Tap “Find My iPhone.” 5. Toggle on “Find My iPhone,” “Enable Offline Finding,” and “Send Last Location.”

If your iPhone is ever lost or stolen, you can: 1. Go to iCloud.com or use the Find My app on another Apple device. 2. Sign in with your Apple ID. 3. Select your missing device from the list. 4. Choose to play a sound, lock the device with a passcode, display a message, or erase all data.

For example, if you misplace your iPhone at a coffee shop, you can use the Find My app on a friend’s device to play a sound, helping you locate it quickly. If it was stolen, you can lock it remotely to prevent access and display a contact message for anyone who finds it. If recovery is unlikely, you can erase your data to protect your privacy.

Accessibility

Tip 76: Use The Guided Access

Guided Access helps you stay focused on a single app by temporarily restricting your iPhone to that app and disabling certain areas of the screen. This feature is especially useful when you want to limit distractions or prevent accidental taps during presentations, exams, or when handing your phone to a child.

Watch Tutorial:
Tap here
https://link.content360.io/Use-The-Guided-Access

Figure 76: Scan to Watch

To enable and use Guided Access: 1. Go to Settings > Accessibility > Guided Access and toggle it on. 2. Tap Passcode Settings to set a Guided Access passcode or enable Face ID/Touch ID for ending sessions. 3. Open the app you want to use. 4. Triple-click the side or home button to start Guided Access. 5. Adjust settings by circling areas of the screen you want to disable or tap Options to restrict hardware buttons, touch, or motion. 6. Tap Start to begin the session. 7. To end Guided Access, triple-click the side or home button again, enter your passcode or use Face ID/Touch ID, and tap End.

Use Guided Access when lending your phone to someone for a specific task, such as showing a photo, playing a game, or using an educational app. It prevents accidental exits or unintended changes, ensuring the device stays locked in the intended app until you decide to exit.

Tip 77: Use Assistive Touch

Assistive Touch is an accessibility feature on the iPhone that provides an on-screen menu for users to access various functions without using physical buttons. It is especially helpful for those with limited motor skills or when hardware buttons are unresponsive. This feature allows you to perform gestures, adjust volume, lock the screen, and more, all through a customizable touch interface.

Watch Tutorial:
Tap here
https://link.content360.io/Use-Assistive-Touch

Figure 77: Scan to Watch

To enable and use Assistive Touch: 1. Open the Settings app on your iPhone. 2. Tap on Accessibility. 3. Select Touch. 4. Tap AssistiveTouch and toggle it on. 5. A floating button will appear on the screen. Tap this button to access the Assistive Touch menu. 6. From the menu, you can access functions such as Home, Siri, Control Center, Notifications, and custom gestures. 7. To customize the menu, tap Customize Top Level Menu within the AssistiveTouch settings and add or remove icons as desired.

Assistive Touch is particularly useful if your iPhone's physical buttons, like the Home or Volume buttons, are damaged or difficult to press. It also aids users who prefer using touch commands instead of hardware controls, making navigation easier and more accessible in various situations.

Tip 78: Use Voice-Over

Voice-Over is an accessibility feature on the iPhone that reads aloud the content on your screen, helping users who have difficulty seeing or reading text. It provides spoken descriptions of everything happening on your device, including notifications, battery status, and text selection, allowing for easier navigation without relying solely on visual cues.

Watch Tutorial:
Tap here
https://link.content360.io/iphone-voiceover-screen-reader

Figure 78: Scan to Watch

To enable and use Voice-Over: 1. Open the Settings app on your iPhone. 2. Tap on "Accessibility." 3. Select "VoiceOver" under the Vision section. 4. Toggle the VoiceOver switch to turn it on. 5. Once activated, tap once to select an item on the screen, then double-tap to activate it. 6. Swipe left or right with one finger to move to the previous or next item. 7. To adjust the speaking rate or voice, return to the VoiceOver settings and customize as desired.

Voice-Over is especially useful for individuals with visual impairments who want to navigate their iPhone independently. It can also assist anyone who prefers hands-free interaction, such as when cooking or multitasking, by audibly guiding them through apps and messages without needing to look at the screen.

Tip 79: Use Switch Control

Switch Control is an accessibility feature on the iPhone that allows users to navigate and interact with their device using adaptive switches or the screen itself. It is designed to assist individuals with limited mobility by enabling them to control the iPhone through simple scanning and selection methods.

Watch Tutorial:
Tap here
https://link.content360.io/iphone-switch-control-accessibility

Figure 79: Scan to Watch

To enable and use Switch Control:

1. Open the Settings app on your iPhone.
2. Tap on Accessibility.
3. Scroll down and select Switch Control.
4. Toggle the Switch Control option to on.
5. Configure switches by selecting "Switches" and adding new switches, which can be assigned to various actions such as screen taps or gestures.
6. Once set up, Switch Control will scan the screen items automatically, highlighting each selectable item.
7. Use your switch to select the highlighted item, allowing you to navigate apps, type, and perform other functions without touching the screen directly.

Switch Control is particularly useful for users with motor impairments who find traditional touch gestures challenging. For example, someone with limited hand movement can use a single adaptive switch to browse through messages, answer calls, or interact with apps, providing greater independence and accessibility.

Tip 80: Use Voice Control

Voice Control allows you to operate your iPhone entirely with your voice, making it easier to navigate, dictate text, and control apps without touching the screen. This feature is especially helpful for individuals with limited mobility or when your hands are occupied.

Watch Tutorial:
Tap here
https://link.content360.io/Use-Voice-Control

Figure 80: Scan to Watch

To enable and use Voice Control: 1. Open the Settings app on your iPhone. 2. Scroll down and tap on "Accessibility." 3. Select "Voice Control." 4. Tap "Set Up Voice Control," then follow the on-screen instructions to download the necessary files. 5. Once set up, toggle the Voice Control switch to turn it on. 6. Use commands like "Open Safari," "Go home," "Scroll down," or dictate messages by saying what you want to type. 7. To see a list of available commands, say "Show commands."

Use Voice Control when you need hands-free operation, such as cooking while following a recipe, exercising, or when your hands are dirty or busy. It enhances accessibility and convenience by allowing you to interact with your iPhone using simple voice commands.

Tip 81: Use Sound Recognition

Sound Recognition is an accessibility feature on the iPhone that listens for specific sounds and notifies you when they are detected. This can be helpful for users who are hard of hearing or in situations where you may not hear important sounds, such as a doorbell, a smoke alarm, or a baby crying.

Watch Tutorial:
Tap here
https://link.content360.io/Use-Sound-Recognition

Figure 81: Scan to Watch

To enable Sound Recognition: 1. Open the Settings app. 2. Tap on Accessibility. 3. Scroll down and select Sound Recognition. 4. Toggle Sound Recognition to On. 5. Tap on Sounds to choose which specific sounds you want your iPhone to listen for, such as Fire Alarm, Doorbell, or Baby Crying. 6. Once set, your iPhone will send an alert when it detects one of the selected sounds.

Use Sound Recognition to stay aware of important auditory cues when you are in a noisy environment, wearing headphones, or have hearing difficulties. For example, if you are working in a busy office, you can set your iPhone to notify you of a smoke alarm or doorbell, ensuring you don't miss critical alerts even if you don't hear them directly.

Tip 82: Use Live Captions

Live Captions automatically generate real-time subtitles for any audio playing on your iPhone, including videos, phone calls, and media apps. This feature enhances accessibility by allowing you to read spoken content when you cannot listen to audio or when the environment is noisy.

Watch Tutorial:
Tap here
https://link.content360.io/Use-Live-Captions

Figure 82: Scan to Watch

To enable Live Captions on your iPhone, go to Settings, then tap Accessibility. Scroll down and select Live Captions, then toggle the switch to turn it on. You can adjust the caption size and language preferences within this menu to suit your needs.

Live Captions are especially useful during meetings or lectures when you want to follow along without audio distractions. They also help when watching videos in public places where sound is muted, or when communicating in noisy environments where hearing clearly is difficult.

Tip 83: Use Personal Voice

Personal Voice allows you to create a customized voice that sounds like you, which can be used for VoiceOver, Siri, and other accessibility features. This enhances the personalization and naturalness of interactions with your iPhone, making it easier and more enjoyable to use.

Watch Tutorial:
Tap here
https://link.content360.io/Use-Personal-Voice

Figure 83: Scan to Watch

Steps: 1. Open the Settings app on your iPhone. 2. Scroll down and tap on Accessibility. 3. Select Spoken Content. 4. Tap on Personal Voice. 5. Follow the prompts to record a series of phrases; this process trains your iPhone to replicate your voice. 6. Once the recording is complete and processed, your Personal Voice will be available to use. 7. Enable Personal Voice for supported features like VoiceOver or announcements through the Spoken Content settings.

Use Case: If you rely on VoiceOver for navigating your iPhone, using Personal Voice can make the experience feel more familiar and comfortable by hearing your own voice read text aloud. This can be especially meaningful for users with speech impairments or those who want a more personalized interaction with their device.

Tip 84: Use Point And Speak

Point and Speak is a useful iPhone feature that allows you to quickly capture information by simply pointing your camera at text or an object and having your device read it aloud or recognize it. This can be especially helpful for reading signs, menus, or labels without needing to type or manually search for information.

Watch Tutorial:
Tap here
https://youtube.com/shorts/LN1cLc9NO-o?si=9fXKpSb_MyZehapj

Figure 84: Scan to Watch

To use Point and Speak: 1. Open the Camera app on your iPhone. 2. Point the camera at the text or object you want to capture. 3. Tap the Live Text or text recognition icon that appears on the screen. 4. Select the desired text or information. 5. Choose the Speak option from the menu to have the text read aloud, or use the available actions like copy, look up, or translate.

Use Case: Imagine you are in a foreign country and come across a menu written in a language you don't understand. By using Point and Speak, you can point your iPhone at the menu, have the text recognized, and then use the translate or speak function to understand your options without needing to type anything manually.

Tip 85: Use Assistive Access

Assistive Access is a feature on the iPhone designed to simplify the user interface, making it easier to navigate for individuals who may have difficulty using the standard iPhone layout. It provides a streamlined experience by limiting the number of available apps and options, reducing distractions, and enabling customized accessibility settings tailored to specific needs.

Watch Tutorial:
Tap here
https://link.content360.io/Use-Assistive-Access

Figure 85: Scan to Watch

To enable Assistive Access: 1. Open the Settings app on your iPhone. 2. Scroll down and tap on Accessibility. 3. Select Assistive Access. 4. Toggle on Assistive Access. 5. Customize the options by selecting which apps and features you want available during Assistive Access mode. 6. To activate Assistive Access quickly, you can set up an Accessibility Shortcut by going to Settings > Accessibility > Accessibility Shortcut and selecting Assistive Access.

Use Assistive Access when handing your iPhone to someone who may find the full interface overwhelming, such as children, elderly family members, or individuals with cognitive or motor challenges. It helps prevent accidental taps on unintended apps or settings and focuses the device's functionality on the essential tasks they need to perform.

Tip 86: Use Door Detection

Door Detection is an accessibility feature on the iPhone that helps users who are visually impaired or have difficulty navigating unfamiliar environments. It uses the device's camera and machine learning to identify doors, providing information about their location and distance. This feature can assist users in safely approaching and opening doors.

Watch Tutorial:
Tap here
https://link.content360.io/Use-Door-Detection

Figure 86: Scan to Watch

To enable and use Door Detection:

1. Open the Settings app on your iPhone.
2. Scroll down and tap on Accessibility.
3. Select "Magnifier" from the list.
4. Turn on the Magnifier toggle if it's not already enabled.
5. Scroll down within the Magnifier settings and enable "Door Detection."
6. To use the feature, activate the Magnifier by triple-clicking the side or home button.
7. Point your iPhone's camera towards a door; the device will announce the presence of the door and provide distance information.

Use Door Detection when navigating unfamiliar buildings, such as offices, hotels, or public facilities. It helps users identify door locations and assess how far they are, making it easier to approach and open doors without assistance. This feature enhances independence and confidence when moving through new or complex environments.

Tip 87: Use Background Sounds

Background Sounds on your iPhone help create a more focused and calming environment by playing subtle audio like rain, ocean, or white noise. These sounds can mask distracting noises, aid concentration, or promote relaxation, especially in noisy settings.

Watch Tutorial:
Tap here
https://link.content360.io/Use-Background-Sounds

Figure 87: Scan to Watch

To enable Background Sounds: 1. Open the Settings app. 2. Go to Accessibility. 3. Tap on Audio/Visual. 4. Select Background Sounds. 5. Toggle Background Sounds on. 6. Choose a sound from options like Rain, Ocean, or Balanced Noise. 7. Adjust the volume slider to your preference. 8. Optionally, enable "Use When Media Is Playing" to blend sounds with other audio.

Use Background Sounds when working in a busy café or a bustling home environment to help maintain focus. They can also be useful during meditation or when trying to fall asleep, providing a consistent and soothing audio backdrop that minimizes distractions.

Hidden Features & Power Tips

Tip 88: Use Back Tap For Custom Actions

Back Tap is a feature on the iPhone that allows you to perform custom actions by simply tapping the back of your device. This gesture-based shortcut can be set up to trigger a variety of functions such as opening apps, taking screenshots, or activating accessibility features, providing quick access without navigating through menus.

Watch Tutorial:
Tap here
https://link.content360.io/Use-Back-Tap-For-Custom-Actions

Figure 88: Scan to Watch

To enable and customize Back Tap: 1. Open the Settings app on your iPhone. 2. Scroll down and select Accessibility. 3. Tap on Touch. 4. Scroll to the bottom and choose Back Tap. 5. Select either Double Tap or Triple Tap. 6. Choose the action you want to assign from the list, such as opening Control Center, launching an app, or running a shortcut.

For example, if you frequently need to take screenshots, you can set a Double Tap on the back of your iPhone to capture the screen instantly. This saves time and effort compared to pressing the physical buttons, especially when your hands are occupied or you're in a hurry.

Tip 89: Use Shake To Undo

Shake To Undo is a convenient feature on the iPhone that allows you to quickly undo your last action, such as typing or deleting text, by simply shaking your device. Instead of manually tapping the undo button, this gesture helps you correct mistakes faster and improves your overall typing experience.

Watch Tutorial:
Tap here
https://link.content360.io/Use-Shake-To-Undo

Figure 89: Scan to Watch

To use Shake To Undo:

1. Type or delete text in any app that supports text input, like Messages, Notes, or Mail.
2. If you want to undo your last action, give your iPhone a gentle shake.
3. A popup will appear asking if you want to undo the last change. Tap "Undo" to revert the action or "Cancel" to keep it.
4. If you want to redo the action, you can shake again and tap "Redo."

Imagine you are composing a message and accidentally delete an important sentence. Instead of retyping everything, you just shake your iPhone, tap Undo, and your deleted text instantly reappears, saving you time and effort.

Tip 90: Use The Hidden Trackpad

The iPhone's keyboard includes a hidden trackpad feature that makes cursor control easier and more precise. Instead of tapping repeatedly to position the cursor, you can turn the keyboard into a trackpad by pressing and holding on the space bar. This allows you to move the cursor smoothly around the text, making editing faster and more accurate.

Watch Tutorial:
Tap here
https://link.content360.io/Use-The-Hidden-Trackpad

Figure 90: Scan to Watch

Steps:
1. Open any app where you can type, such as Messages or Notes.
2. Tap the text field to bring up the keyboard.
3. Press and hold the space bar until the keys disappear and the keyboard turns gray.
4. Without lifting your finger, slide it around to move the cursor to the desired position.
5. Release your finger to place the cursor.

Use Case:
When editing a long message or document, positioning the cursor precisely can be challenging. Instead of tapping multiple times or holding down the screen to magnify the text, use the hidden trackpad. This method lets you glide the cursor to the exact spot for quick corrections, saving time and reducing frustration.

Tip 91: Use The Scientific Calculator

The iPhone's Calculator app includes a hidden scientific calculator that appears when you rotate your device to landscape mode. This feature provides additional functions such as trigonometric calculations, exponents, logarithms, and more, making it useful for advanced math tasks without needing to download a separate app.

Watch Tutorial:
Tap here
https://link.content360.io/Use-The-Scientific-Calculator

Figure 91: Scan to Watch

To access the scientific calculator: 1. Open the Calculator app on your iPhone. 2. Hold your phone in portrait orientation to see the standard calculator. 3. Rotate your iPhone sideways to landscape orientation. 4. The calculator will automatically transform into the scientific version, displaying extra buttons for advanced functions.

Use this feature when you need to perform more complex calculations, such as solving trigonometric problems for a math class, calculating logarithms for a science project, or handling exponential functions in financial planning. It saves time by providing powerful tools within the default app without switching to third-party software.

Tip 92: Use Custom Vibrations For Contacts

Custom vibrations allow you to identify who is calling or messaging without looking at your iPhone. By assigning unique vibration patterns to specific contacts, you can recognize them through touch alone, which is especially useful in situations where sound or visual alerts are impractical.

Watch Tutorial:
Tap here
https://link.content360.io/Use-Custom-Vibrations-For-Contacts

Figure 92: Scan to Watch

Steps: 1. Open the Contacts app and select the contact you want to assign a custom vibration to. 2. Tap “Edit” in the upper right corner. 3. Scroll down and tap on “Ringtone” or “Text Tone,” depending on whether you want to customize calls or messages. 4. Tap “Vibration” at the top of the screen. 5. Choose a predefined vibration pattern or tap “Create New Vibration” to record a custom pattern. 6. Tap “Save” to assign the vibration to that contact. 7. Tap “Done” to finish editing the contact.

Use Case: When you’re in a meeting or a noisy environment where you cannot check your phone visually or audibly, custom vibrations help you instantly know who is contacting you. For example, assigning a unique vibration to your boss allows you to prioritize their calls or messages without interrupting others or pulling out your phone.

Tip 93: Use The Hidden Spirit Level

The iPhone includes a hidden spirit level within the Measure app, which helps you check if a surface is perfectly horizontal or vertical. This tool is useful for tasks that require precision, such as hanging pictures, aligning furniture, or leveling objects.

Watch Tutorial:
Tap here
https://link.content360.io/Use-The-Hidden-Spirit-Level

Figure 93: Scan to Watch

To access and use the hidden spirit level:

1. Open the Measure app on your iPhone.
2. Tap the "Level" option at the bottom of the screen.
3. Place your iPhone on the surface you want to check.
4. Observe the angle readings and the visual indicators on the screen.
5. Adjust the object or surface until the readings show zero degrees, indicating it is level.

For example, when hanging a framed photo, you can place your iPhone on top of the frame and adjust it until the spirit level shows that it is perfectly horizontal. This ensures your picture hangs straight without needing a separate tool.

Tip 94: Use Silence Unknown Callers

Silence Unknown Callers is a feature on the iPhone that automatically silences calls from numbers that are not in your contacts, recent calls, or Siri Suggestions. When enabled, these calls are sent directly to voicemail and appear in the Recent Calls list without ringing or vibrating, reducing interruptions from spam or unwanted calls while still allowing important contacts to reach you.

Watch Tutorial:
Tap here
https://link.content360.io/Use-Silence-Unknown-Callers

Figure 94: Scan to Watch

To enable Silence Unknown Callers: 1. Open the Settings app on your iPhone. 2. Scroll down and tap on Phone. 3. Find and select Silence Unknown Callers. 4. Toggle the switch to turn the feature on.

This feature is particularly useful when you want to minimize distractions during work hours or personal time without blocking numbers outright. It helps filter out potential spam calls while ensuring calls from known contacts and important numbers still come through. For example, if you are in a meeting or focusing on a project, enabling Silence Unknown Callers allows you to avoid interruptions from unknown numbers while still staying accessible to colleagues and friends.

Tip 95: Use Wi-Fi Password Sharing

Wi-Fi password sharing allows you to quickly and securely share your saved Wi-Fi network password with nearby Apple devices without revealing the actual password. This feature simplifies connecting guests to your network without the hassle of typing long passwords.

Watch Tutorial:
Tap here
https://link.content360.io/Use-Wi-Fi-Password-Sharing

Figure 95: Scan to Watch

Steps:
1. Ensure both devices have Wi-Fi and Bluetooth turned on and are within close proximity.
2. The device requesting the Wi-Fi connection should select the network from the list of available Wi-Fi networks.
3. On your iPhone (the one already connected to the Wi-Fi), a prompt will appear asking if you want to share the password.
4. Tap "Share Password" on your iPhone.
5. The other device will automatically receive the password and connect to the Wi-Fi network.
6. After the connection is successful, tap "Done" on your iPhone.

Use Case:
When friends or family visit your home and want to join your Wi-Fi network, you can instantly share the password without having to spell it out or type it manually. This saves time and reduces errors, especially for complex passwords, making it a convenient way to keep everyone connected.

Tip 96: Use The Built-In QR Scanner

The iPhone has a built-in QR code scanner integrated directly into the Camera app, allowing you to quickly scan QR codes without needing a separate app. When you point your camera at a QR code, the phone automatically recognizes it and displays a notification with the relevant action, such as opening a website, adding a contact, or connecting to a Wi-Fi network.

Watch Tutorial:
Tap here
https://link.content360.io/Use-The-Built-In-QR-Scanner

Figure 96: Scan to Watch

Steps:
1. Open the Camera app on your iPhone.
2. Point the camera steadily at the QR code you want to scan.
3. Wait for a notification banner to appear at the top of the screen.
4. Tap the notification to perform the action associated with the QR code.

Use Case:
Imagine you're at a restaurant that has a QR code on the table for viewing the menu. Instead of downloading an app or typing in a website, you simply open your iPhone's Camera app, point it at the QR code, and tap the notification that appears. This instantly opens the menu on your browser, making it quick and hassle-free to check the options.

Tip 97: Use Live Activities

Live Activities allow you to stay updated with real-time events directly on your iPhone's Lock Screen or Dynamic Island without needing to open an app. This feature provides ongoing information such as sports scores, delivery statuses, or workout progress, making it easier to keep track of important updates at a glance.

Watch Tutorial:
Tap here
https://link.content360.io/Use-Live-Activities

Figure 97: Scan to Watch

Steps:
1. Ensure your iPhone is updated to iOS 16.1 or later, as Live Activities require this version or newer.
2. Open an app that supports Live Activities, such as a sports app or a food delivery app.
3. Start an activity within the app; for example, begin tracking a live game or place a food order.
4. The Live Activity will appear on your Lock Screen or Dynamic Island, showing real-time updates.
5. To interact with the Live Activity, tap it to open the corresponding app for more details.
6. To stop the Live Activity, either let it end naturally or tap and hold the Live Activity, then select "End" or close it from the app.

Use Case
When ordering food for delivery, Live Activities keep you informed on the preparation and delivery status without unlocking your phone or opening the app repeatedly. You can glance at your Lock Screen to see when your order is being prepared, picked up, or arriving, saving time and keeping you informed effortlessly.

Tip 98: Use StandBy Mode

StandBy Mode transforms your iPhone into a useful smart display when it is charging and placed in landscape orientation. It shows glanceable information such as the time, calendar events, weather, and smart home controls, making it perfect for use on a nightstand or desk.

Watch Tutorial:
Tap here
https://link.content360.io/Use-StandBy-Mode

Figure 98: Scan to Watch

Steps
1. Connect your iPhone to a charger.
2. Place the iPhone on its side in landscape orientation.
3. The StandBy Mode interface will automatically appear, displaying widgets and information.
4. Customize the displayed content by going to Settings > StandBy, where you can select preferred widgets and styles.

Use Case
When charging your iPhone overnight on a bedside table, StandBy Mode lets you see the time, upcoming alarms, and notifications at a glance without unlocking the phone. It also enables quick access to smart home controls, such as adjusting lights or thermostat settings, creating a convenient and hands-free experience throughout the day or night.

Tip 99: Use Action Button

The Action Button on the latest iPhone models provides quick access to customizable functions, allowing you to perform tasks instantly without navigating through multiple menus. By assigning a specific action to the button, you can streamline your workflow and enhance device usability.

Watch Tutorial:
Tap here
https://link.content360.io/Use-Action-Button

Figure 99: Scan to Watch

Steps:
1. Open the Settings app on your iPhone.
2. Scroll down and select "Action Button."
3. Choose your preferred function from the list, such as launching the camera, turning on the flashlight, opening a specific app, or starting a voice memo.
4. Once selected, press the Action Button to perform the assigned task instantly.

Use Case:
If you frequently use the voice memo app to record ideas or meetings, assigning it to the Action Button allows you to start a recording with a single press, eliminating the need to unlock your phone and search for the app. This quick access saves time and ensures you never miss capturing important audio.

Tip 100: Use Journal App

The Journal app on your iPhone helps you capture daily thoughts, experiences, and ideas in an organized and easily accessible way. It encourages consistent reflection and can improve mindfulness, creativity, and productivity by providing a dedicated space for personal entries.

Watch Tutorial:
Tap here
https://link.content360.io/Use-Journal-App

Figure 100: Scan to Watch

To use the Journal app effectively:
1. Open the Journal app on your iPhone.
2. Tap the "+" or "New Entry" button to start a new journal entry.
3. Write about your thoughts, events, or ideas; you can add photos, tags, or locations to enrich your entry.
4. Use prompts provided by the app if you need inspiration or want to focus on specific topics.
5. Save the entry and revisit past entries anytime to track your progress or reflect on experiences.

For example, you might use the Journal app each evening to document the highlights of your day, jot down lessons learned, or set intentions for tomorrow. This practice can help reduce stress, increase gratitude, and keep personal growth on track.

Learn More

Figure 101: Commercial

Special Thanks

This book was made possible thanks to the tools, platforms, and creators who continue to push innovation forward.

Special thanks to:

- Content creators and educators
- Apple
- Developers and engineers
- AI tools
- Open-source communities

And most importantly, to you—the reader.

References

- https://link.content360.io/iphone-3d-touch-quick-actions
- https://link.content360.io/iphone-reachability-one-hand
- https://link.content360.io/iphone-magnifier-accessibility-tool
- https://link.content360.io/iphone-calculator-hidden-gestures
- https://link.content360.io/spotlight_search
- https://link.content360.io/back_tap
- https://link.content360.io/customize_home_screen
- https://link.content360.io/use_widgets_lock
- https://link.content360.io/setup_face_id
- https://link.content360.io/app_library
- https://link.content360.io/siri_shortcut
- https://link.content360.io/use_quick_settings
- https://link.content360.io/use_haptic_touch
- https://link.content360.io/iphone-flashlight
- https://link.content360.io/low_power_mode
- https://link.content360.io/use_do_not_disturb
- https://link.content360.io/iphone-night-shift-display
- https://link.content360.io/set_timer
- https://link.content360.io/customize_control_center
- https://link.content360.io/Control_center_settings
- https://link.content360.io/use_screen_time
- https://link.content360.io/wallet_app
- https://link.content360.io/use_apple_pay
- https://link.content360.io/weather_app
- https://link.content360.io/use_focus_app
- https://link.content360.io/Optimize_battery_charging
- https://link.content360.io/iphone-camera-self-timer
- https://link.content360.io/iphone-camera-burst-mode
- https://link.content360.io/iPhone-camera-grid
- https://link.content360.io/iphone-camera-level
- https://link.content360.io/iphone-live-photos
- https://link.content360.io/use_portrait_mode
- https://link.content360.io/cinematic_mode
- https://link.content360.io/cut_out_mode
- https://link.content360.io/Visual_look_up
- https://link.content360.io/Use_live_text
- https://link.content360.io/use_photo_app
- https://link.content360.io/Use-ProRAW-ProRes
- https://link.content360.io/Use-Photographic-Styles
- https://link.content360.io/iphone-facetime-video-calls
- https://link.content360.io/iphone-imessage-features

- https://link.content360.io/iphone-airdrop-file-sharing
- https://link.content360.io/iphone-notes-organization
- https://link.content360.io/iphone-podcasts-listening
- https://link.content360.io/iphone-apple-tv-streaming
- https://link.content360.io/Use-SharePlay
- https://link.content360.io/Use-NameDrop
- https://link.content360.io/Use-Check-In-Safety-Feature
- https://link.content360.io/Use-Live-Voicemail
- https://link.content360.io/Use-Contact-Posters
- https://link.content360.io/Use-Stickers-From-Photos
- https://link.content360.io/Use-The-Measure-App
- https://link.content360.io/Use-The-Files-App
- https://link.content360.io/Use-The-Shortcuts-Widget
- https://link.content360.io/Use-Car-Play
- https://link.content360.io/iphone-screen-mirroring
- https://link.content360.io/iPhone-Screen-Recording
- https://link.content360.io/iphone-icloud-backup-sync
- https://link.content360.io/iphone-family-sharing-setup
- https://link.content360.io/Use-Drag-And-Drop-Between-Apps
- https://link.content360.io/Use-Text-Replacement-Shortcuts
- https://link.content360.io/Use-Scan-Documents-With-Notes
- https://link.content360.io/Use-Handoff-Between-Devices
- https://link.content360.io/Use-Emergency-SOS
- https://link.content360.io/iphone-find-my-device-tracking
- https://link.content360.io/Use-Lockdown-Mode
- https://link.content360.io/Use-Safety-Check
- https://link.content360.io/Use-App-Privacy-Report
- https://link.content360.io/Use-Mail-Privacy-Protection
- https://link.content360.io/Use-Hide-My-Email
- https://link.content360.io/Use-Private-Browsing-In-Safari
- https://link.content360.io/Use-Passkeys-Instead-Of-Passwords
- https://youtu.be/8NrfcyA3Bf8?si=dLIPA2-KrkU9YJ26
- https://link.content360.io/Use-Crash-Detection
- https://link.content360.io/Use-Stolen-Device-Protection
- https://link.content360.io/Use-The-Guided-Access
- https://link.content360.io/Use-Assistive-Touch
- https://link.content360.io/iphone-voiceover-screen-reader
- https://link.content360.io/iphone-switch-control-accessibility
- https://link.content360.io/Use-Voice-Control
- https://link.content360.io/Use-Sound-Recognition
- https://link.content360.io/Use-Live-Captions
- https://link.content360.io/Use-Personal-Voice
- https://youtube.com/shorts/LN1cLc9NO-o?si=9fXKpSb_MyZehap

- https://link.content360.io/Use-Assistive-Access
- https://link.content360.io/Use-Door-Detection
- https://link.content360.io/Use-Background-Sounds
- https://link.content360.io/Use-Back-Tap-For-Custom-Actions
- https://link.content360.io/Use-Shake-To-Undo
- https://link.content360.io/Use-The-Hidden-Trackpad
- https://link.content360.io/Use-The-Scientific-Calculator
- https://link.content360.io/Use-Custom-Vibrations-For-Contacts
- https://link.content360.io/Use-The-Hidden-Spirit-Level
- https://link.content360.io/Use-Silence-Unknown-Callers
- https://link.content360.io/Use-Wi-Fi-Password-Sharing
- https://link.content360.io/Use-The-Built-In-QR-Scanner
- https://link.content360.io/Use-Live-Activities
- https://link.content360.io/Use-StandBy-Mode
- https://link.content360.io/Use-Action-Button
- https://link.content360.io/Use-Journal-App

Made in the USA
Monee, IL
06 November 2026

67383113R00063